Astronomers' Universe

For other titles published in this series, go to
http://www.springer.com/series/6960

Gerhard Börner

The Wondrous Universe

Creation without Creator?

 Springer

Prof. Dr. Gerhard Börner
MPI für Astrophysik
Karl-Schwarzschild-Str. 1
85740 Garching
Germany
grb@MPA-Garching.MPG.DE

ISSN 1614-659X
ISBN 978-3-642-20103-5   e-ISBN 978-3-642-20104-2
DOI 10.1007/978-3-642-20104-2
Springer Heidelberg Dordrecht London New York

Library of Congress Control Number: 2011934435

Cover design: eStudio Calamar S.L.

Cover figure: Pillars of Creation – Gas Pillars in a Star-Forming Region in the Eagle Nebula (M16).
           Credit: NASA, ESA, STScI, J. Hester and P. Scowen (Arizona State University)

Printed on acid-free paper

Springer is part of Springer Science+Business Media (www.springer.com)

Contents

1. The Wondrous Universe

1.1 The Starry Heavens

The stars began to shine unusually bright and clear above the observatory in the Arizona desert. Gradually the broad luminous band of the Milky Way appeared distinctly. The astronomers had ended their day's sleep, and now went to work carrying liquid nitrogen to cool their instruments, checking the CCD pixels in the twilight, and then pointing their telescopes toward faraway galaxies and star clusters. As a short-term visitor, I had the opportunity to look over their shoulders at the images of distant galaxies which were transmitted from the telescope to a computer screen.

The small roads between the observatory domes were un-lighted to avoid disturbing strong radiation. While driving from one observing station to the next, I began to dreamily think about stars, galaxies, the whole universe, and its relation to mankind.

I felt awe and wonder regarding the vast extent of the universe with its wonderful variety of stars, the many shapes and great number of galaxies. Every star has its own interesting appearance, history, and fate – most shine quietly like our Sun, some swell up to become red giants, others dwindle to the size of white dwarfs, while quite a few blow up as a supernova and end as neutron stars and black holes. We have our home in this immense and amazing cosmos, and we can even understand its workings.

Almost everybody, I guess, has experienced the mystery and beauty of the sky at night. "The starry heavens inside me, and the moral law above me..." – this slightly changed quotation from Immanuel Kant has been used by the philosopher and physicist Carl-Friedrich von Weizsäcker, when he described his feelings as a 12-year-old boy, who studied the stars intently with the help of a

G. Börner, *The Wondrous Universe*, Astronomers' Universe,
DOI 10.1007/978-3-642-20104-2_1,
© Springer-Verlag Berlin Heidelberg 2011

sky atlas (Kant wrote: "...The starry heavens above, and the moral law inside me").

What can we learn from looking at the stars?

Since ancient times humans have speculated about the nature of the universe and man's place in it. The obvious facts that the Sun rises and sets regularly, that the Moon varies in shape, and that the planets move across the sky led to the view that the Earth was at the center of the universe, while the Sun and the planets moved around it bound to celestial spheres. More detailed observations required that this geocentric world view had to be modified. Since the ancient astronomers were convinced that all motions in the sky must be circular, they were forced to invent a complex system of cycles and epicycles, i.e., circles rolling along circles, to accommodate all the observations. Then the "Ptolemaic" world view was abandoned in favor of a heliocentric system which seemed more natural to Copernicus, and which made it easier for Johannes Kepler to fit the precise data obtained by Tycho Brahe. He proposed a system, where the Earth and the planets move on elliptical orbits around the Sun. Kepler's laws of planetary motion found an explanation in Isaac Newton's law of gravitation. Newton has shown us how the same law of force between two masses can describe the fall of an apple from its tree, as well as the motion of the planets around the Sun. As a consequence of this gain in knowledge, man was decisively removed from his position at the center of the universe.

With the new telescopes many more objects were discovered in the heavens, and mankind's place in the cosmos moved away even further from a central position to the surface of a small planet orbiting a commonplace star, one among billions of its kind. At the beginning of the twentieth century such was the generally accepted view of the world among astronomers: The cosmos was an inverse assembly of stars extending on without end, unchanging, and uniform.

But this view underwent a radical change, when mainly through the work of the American astronomer Edwin Hubble it was discovered that the stars were arranged in "island universes," huge systems containing billions of stars, among them our own "galaxy," the Milky Way, and that all those stellar systems were apparently moving away from each other at high speed. Obviously the view of a static world had to be abandoned, and the unchanging starry heavens had to be replaced by a dynamic cosmos which had evolved to its present state from a beginning quite different from what we see now. The argument for a universe in evolution has been strengthened by many subsequent observations, and also by the cosmological models derived from Albert Einstein's theory of gravitation.

All that we shall discuss in detail in the following section. In a very preliminary way, we can say that by looking at the stars we have arrived at an astonishing insight: We have found the complex structure of the universe, its origin in a hypothetical big bang, and its subsequent evolution from a diffuse hot gas to a complex system of galaxies. Such a picture exceeds our everyday experience by far, and even appears to be in conflict with common sense at certain points. Who could have imagined a world in which even space and time originate with the big bang, and pass away in black holes? It thus seems that space and time are subject to change and can no longer be regarded as absolute categories for our experience. Maybe our existence in space and time is only one partial aspect of reality?

Certainly the old questions "Where do we come from?" and "Whereto are we going?" can now be asked not only in a biological, but also in a cosmological context. Modern cosmology, together with modern biology, tells us that we are the result of a long chain of evolution in a cosmos, where not a single atom is lost and where our life is even connected to the evolution of the stars.

1.2 Particles and Fields

When we turn our eyes away from the stars and look at the everyday world around us, we are confronted by the astonishing discoveries made by the physicists in the world of atoms and subatomic elementary particles. They show us a microcosm supporting our solid normal world by a subsoil of particles and fields. Strangely, these objects behave in a way that makes it impossible to describe their properties in terms of our everyday classical world: Particles are also waves, and waves are particles, and they appear as one or the other depending on how we look at them.

Before we dive deeply into this "quantum world" there can be no harm in keeping the fact in view that many things, including some in common, everyday use, are quite strange. Light for example, radio- and microwaves, and X-rays – all are electromagnetic waves distinguished only by a different wavelength. According to quantum mechanics these waves can also be viewed as a stream of particles of a certain energy, but with zero mass. These radiation particles, the "photons," show their presence most directly and clearly in the "photokinetic" effect which has been explained by Albert Einstein in 1905: When a metal foil is irradiated by light, an electric current is generated. This is due to the photons kicking out electrons from their bound state in the atomic lattice, and thus setting them free to move around as an electric current. But this well-known electromagnetic radiation is not a flow of material particles or a vibration of a material system, like a vibrating string. The radiation acts on matter and excites the electrons to oscillations, but it is itself immaterial; the electromagnetic wave determines the possible actions on charged particles. It can be received and emitted as a well-defined form and pattern, but it needs no medium for its propagation.

In the nineteenth century such a behavior was deemed impossible, and therefore the existence of an "ether" was postulated, a ubiquitous substance in which the electromagnetic waves could propagate like waves in water. Albert Einstein then formulated his theory of Special Relativity showing that it was not necessary to have an ether. Electromagnetic waves propagate in empty space without any material substrate. They exist as pure form, immaterial, but nevertheless real objects of the physical world. We cannot really "understand" such objects by reducing the phenomena to a mechanistic picture, where everything happens by position changes of small particles. We just have to get used to it. Physicists speak of the electromagnetic "field," when they designate this object.

It is even more difficult to give an illustration of the subatomic world of elementary particles, where material particles change their identity, can be created from energy, or annihilated into radiation. The theorists try to cope with these phenomena by describing the particles as excitations of fundamental fields. A complete change of the materialistic world view happens here: Matter is not fundamental, but changeable and transient, while the fundamental objects are much less tangible fields. It seems, as if the solid classical world crumbles beneath the pens of theorists. Deep inside, the world seems to be held together not by some touchable material substance, but rather by a mathematical-abstract principle.

This sounds strange indeed, but the strangeness of the quantum objects is not just a vague idea or speculation. Physicists have found in their experiments that the quantum world is really different from the macroscopic, classical one. A famous example is the passage of electrons through two slits in a metal screen. As we shall discuss in some detail later on, the electrons behave either like small particles, whose hits in a detector behind the screen just add up as expected, or as waves, which show a distinct interference pattern. It is amazing that electrons can act as particles or as waves, whereas our classical concepts of particle (compact, confined to a small spatial volume) and wave (extended over a large spatial region showing interference phenomena) seem to be mutually exclusive. The electrons in the double-slit experiment are even much stranger: We can arrange the experiment such that for each electron we can in principle check through which of the two slits it passes. If we carry out this registration, the electrons behave as particles, if we decide not to register the passage, the interference pattern of the "electron waves" appears. The electrons appear as particles or as waves respectively depending on whether they are observed or not observed in their passage of one of the slits.

Obviously quantum objects like electrons behave differently from classical particles. They seem to possess a certain freedom of decision, a quality which breaks up the absolute determinism of the classical world, where one state follows another in strict causality.

This is a very difficult subject which I want to follow up a bit more in Chap. 3 of the book. At this point I can only say that a consistent mathematical formulation of quantum mechanics exists which can deal with such problems for all practical purposes: Quantum mechanics is a well-established and mathematically well-defined structure allowing the computation of atomic processes and properties in perfect agreement with experiments.

Perhaps I can venture a few remarks which may give us some general idea on how this mathematical framework is built up. A quantum mechanical system is described by an assembly of all possible configurations, called states. An electron, for example, may have its spin (its internal rotation) oriented in any particular direction. All these cases are "states" of the system of one spinning electron. The whole system, called "wave function" by the physicists follows a definite time evolution determined by the laws of nature. A specific experiment measures the system, and finds it in one of the possible states; the outcome of the measurement corresponds to a completely random choice among the set of states. The so-called collapse of the wave function is a prescription which works very well in calculations, but it is difficult to make sense of it.

The correct interpretation of quantum mechanics is still the subject of lively debates. There are basically three different approaches, each one not really satisfactory:

The suggestion that there are strictly causal and deterministic processes which produce the quantum phenomena by statistical fluctuations seemed to be an attractive possibility to several eminent physicists, including Albert Einstein. But, as we shall see in Chap. 3, this way of interpretation has been blocked by recent experiments which demonstrate that quantum mechanics cannot be explained by such arguments. The second, very extreme interpretation is the "many-worlds" hypothesis according to which each quantum mechanical event splits the world into a number of parallel universes, one universe for each possible outcome of the measuring process. At each moment the world splits into many new branches and new universes originate in huge numbers. Quite an expensive way to understand quantum mechanics, and too bizarre in my opinion to be taken seriously.

What is left is the so-called Copenhagen interpretation proposed by the founder of quantum mechanics, the Danish physicist Niels Bohr, and his collaborators. Can this interpretation still be considered valid? It basically states that the result of an experiment does not become real until a conscious observer recognizes it. Up to that moment the system floats in a curious "in-between" region, where the wave function contains all possible states. Only when an observer looks at the experiment, one of the states is singled out. This interpretation is considered doubtful by some, because the dividing line between the observer and the quantum mechanical object is not easy to draw.

The situation is loosely speaking as follows: The quantum system produces a signal in some measuring device such as a specific read-out of an index needle. This result is not registered and real, however, until an observer notices it, since we may include the read-out device in the quantum system. And we may continue that way: the glasses of the observer, the excitation in his optical nerve, and so on, maybe up to the moment when a sensory center in the brain of the observer is activated, such that the real separation occurs only through an act of consciousness of the observer. This aspect of the Copenhagen interpretation seems to be in clear contradiction to the idea of an objective real world which can be explored by experiments, and therefore many physicists are motivated to look for other interpretations. We shall discuss this further in Chap. 3 of the book. Do we find here a real dilemma, since the division into subject and objective world becomes impossible, or is this a hint that we must modify quantum mechanics itself?

I do not believe that at this point a boundary appears which in some sense brings transcendent qualities, consciousness, or mind into the world. Some scientists, however, believe just that. They suppose that by including the observer the quantum world acquires a "wholeness" which surpasses the view of a purely objective world. According to their belief quantum mechanics exhibits a basic principle of how the world is constructed differently from the concept of self-organizing matter suggested by classical physics: The quantum world becomes real only when it is reflected in the conscience of an observer. What kind of conscience is necessary? Was there no real world before conscious observers appeared? This would be a truly extreme point of view.

It seems prudent to reserve our judgment on this point, and not to jump to the conclusion that the realistic view of the world is transcended by quantum mechanics, much less to derive a proof for the existence of God from these facts. Nevertheless we should note that the quantum world is different from our well-known classical world, because the strong determinism which has to be obeyed by a system of classical particles does not hold any more. A quantum system is free to choose between several possibilities, and chance plays a role in its development. More freedom than a random choice between different configurations is, of course, not possible for a system of electrons, atoms, and electric fields.

1.3 Physics and Religion

Despite gaps in our knowledge, we find ourselves in a world which we can largely explain in scientific terms. That does not expel its charm. Quite the contrary: The more we understand, the more we look in wonder at the universe. How far does our scientific understanding reach? This question will be followed in this book, and it will be shown how fascinating the images and ideas can be even when confined to physics and biology.

From the electrons dancing in the atom to the stars swinging round the center of the galaxy, everything seems to work in harmony finally creating us on the small and cosmologically insignificant planet Earth.

But is that all? Are we nothing but a product of intricate physical, chemical, and biological processes, an accidental event in the cosmic play which could equally well have not occurred? Our feelings make it difficult for us to answer "yes" to that question. We would like to find some deeper meaning of our existence, and of the occurrence of the universe.

But here the natural sciences remain silent. For good reasons, because the scientific method is restricted to questions of a certain kind. Physicists, for example, ask "How does a body move under the influence of a gravitational field?" or "How is the mass of an elementary particle related to the vacuum expectation value of the Higgs field?" They do not ask "What is gravity?" or "Is the cosmic evolution aimed at producing intelligent beings?" I take these examples from physics, because it is the basic science. All the chemical and biological processes are basically determined by physical laws. Thus in the natural sciences one is trying to find connections in the world with the final aim to reduce everything that can be explored to physical processes. Scientists rely on mathematical formulations of their theoretical concepts and on experiments, and they have built a tightly woven texture of reliable knowledge. Within its boundaries natural science decides on what is true or false, and knowledge gained from science cannot be put in doubt by assertions ignoring scientific results. Science tells us what we can know, but we should not forget its limitations. The evolution in time of a system of atoms or the orbit of a charged particle in a magnetic field is a phenomenon completely free of any meaning. Similarly the evolution of the cosmos is just the history and future of a huge assembly of atoms and fields. But there are many questions of interest to an inquisitive mind to which no answer can be found in the laboratory or observatory. Has the universe any unity or purpose? Is it evolving toward some goal? How did it originate? Is the appearance of life an unimportant random accident in the cosmic history, or does it have a big significance? When we consider the universe moving toward an end, where all life ceases, can we still see value in this cosmic spectacle?

It seems impossible to find answers to such questions except from our inherited religious and ethical conceptions. But what then is the relation to science? This is a difficult matter, and generally we deal with it by carefully separating the fields of science and of religion and philosophy.

But there are overlaps, as in the biblical story of the creation of the world. Although the biblical account is not meant to be a scientific explanation, we may ask in what sense the statement that God has created the world can be understood today. Can the whole creation exist without a creator? Science cannot answer that question, but I hope that by exploring the limits of validity of the scientific description of the world we may open a path to understanding the significance of declarations of faith. Let me mention one example:

We learn from physics that space and time are not given as unchanging properties of reality such as our everyday experience lets us believe. Space and time originate in the big bang and perish at the end-point of the gravitational collapse of large masses in black holes. Black holes and the big-bang origin of the universe are regimes, where our best physical theories are not sufficient to describe what is going on. A more fundamental theory must probably encompass concepts that reach beyond space and time. This also means that we have to consider the possibility that there may exist something not in space and time, something inaccessible to our experience.

These may be just the frontiers of our present scientific knowledge, but perhaps they mark a fundamental final point of the objective, physical explanation of the world. At the moment we do not know, but it need not remain that way. Being an optimist, I expect many new and deep insights from scientific research.

Such insights derived from physics can serve as a solid basis for thinking about the meaning of the cosmic play. Whether we answer "yes" or "no" to the question of "Creation without a creator?" cannot be decided by science, of course, but must remain the personal decision of each individual.

"It is forbidden to mix values and knowledge" said Jacques Monod, one of the founders of molecular biology, and many natural scientists have taken his sentence to heart, and retired moderately to their special research field. But there must be a deep interrelation between both realms, because on the one hand science tries to explain our culture as the outcome of a cosmic evolutionary process, and on the other hand the scientific view of the world belongs inextricably to our culture.

The way we see ourselves is certainly influenced by Darwin's theory of evolution, to name but one example. And the motivation to explore nature, to strive for insights, to choose certain topics of research over others is surely formed by culture and religion.

The scientific method ensured that a densely knit network of secure knowledge was constructed. This is one window open to reality, but is it the only one?

You can adhere to the opinion that everything existing is determined by physics and biology. Everything can be explained by physics arguments, even though there are connections which we have not yet found out. This point of view, however, cannot be established within the system of physics. It is an expression of faith inaccessible to scientific arguments.

Neither can the religious belief be refuted that the world is God's creation. An omnipotent creator can arrange the world just as the physicists find it with properties suggesting a purely physical origin.

Both opinions are viable. It is also possible that science and religion exist in harmony besides each other, as long as the scientists stay within their boundaries, contently doing research on their special topic, and as long as they refrain from postulating a purely biological–physical view of the world for everybody. This is a widespread attitude at present. It resembles a peace between two castles with the drawbridges drawn up, where no real dialogue between theologians and scientists is risked.

But one must not underestimate the attractive power lying in the secure knowledge of science, and the seductive images drawn by scientists for the origin and evolution of the world. Does it not seem, as if everything might be explained, as if the hypothesis of a God as creator of the world might be superfluous, even though it could not be refuted?

Stephen Hawking argues in this sense in his new book – as well as in other books before – that an act of creation is not necessary, because the universe can come into existence on its own. But even if we accepted his mathematical considerations as true, we would only have found a model explaining how the universe evolved from a preexisting quantum structure and not why it has come into being. And then again there is no answer to the question, where the preexisting quantum cosmos comes from.

There are points of contact between science and religion which may be worthwhile aspects of a serious dialogue. Both try to describe the same reality, but from different perspectives, and both use a different language. The aim of a dialogue must therefore at first be a clarification of the conceptions used in their respective regime of validity.

The past few years have seen especially physicists and cosmologists attempting to formulate a uniform view of the world motivated by their claim to propose a theory of everything, and without fear of transgressing the self-imposed limitations of physics.

My opinion is that these attempts should be considered as an interesting impetus for further deliberations. They cannot be the conclusive wisdom, because at present the laws of physics have not yet reached completion in an all-encompassing world formula.

Can we really succeed in a complete description of the world by reducing everything recognizable to physical processes? The experiment is still going on, and the outcome has not been decided. Many, in fact, hold the opinion that the uncertainty of quantum physics or the fine-tuning of the constants of nature is an indication that the attempt to explain everything by a reduction to objective scientific rules is bound to fail. Even a modest question like "Why is the universe as it is, and not different as it could also be according to the laws of physics?" leads immediately to metaphysical considerations.

Even the obvious fact that there is a world requires an explanation. Can physics find a reason, or do we have to fall back on the hypothesis of a creator? What can be concluded from the evolution of the world, from its origin in a simple big bang to complex structures like the human brain?

These questions will accompany us throughout the book and they will be discussed extensively in Chap. 4.

In this book I attempt to give a nontechnical, understandable account of the world view of a physicist, and to put this picture into perspective with respect to our own self-understanding. To start with, I try to present the basic facts known to us from natural sciences. Then I want to discuss the connections to philosophical and theological questions. I hope that I shall succeed in raising the reader's enthusiasm for the wonders of the universe detected by scientific investigations, and in stimulating her or him to follow own thoughts on the subject.

I can think of no better way to end this introduction than with a quotation from Bertrand Russell's introduction to his famous book *History of Western Philosophy* (Allen & Unwin 1961). Russell writes: "Science tells us what we can know, but what we can know is little, and if we forget how much we cannot know we become insensitive to many things of very great importance. Theology, on the other hand, induces a dogmatic belief that we have knowledge where in fact we have ignorance, and by doing so generates a kind of impertinent insolence toward the universe. Uncertainty, in the presence of vivid hopes and fears, is painful, but must be endured if we wish to live without the support of comforting fairy tales."

2. The World at Large: From the Big Bang to Black Holes

First of all we want to look in detail at some insights of modern cosmology and physics, since we want to lay a solid foundation of facts for the scientific view of the world and not just tell a kind of fairy tale.

2.1 Immediate Experiences: A Play of Thoughts

Let us travel in our imagination away from our home on the Earth, even away from the Earth into outer space. As we move away farther and farther, we find that the familiar outlines of houses and streets become more and more vague. From a height of 10 km we see a colorful map, and from 100 km away the circular edge of the terrestrial sphere comes into view. Oceans, continents, and many clouds dominate the picture. From a distance of 100,000 km we see the Earth floating like a blue sphere in the black sky of outer space (see Fig. 2.1).

Our imagined journey then takes us past the Moon. We reach the neighboring planet Mars, and move on past Jupiter and Saturn.

Looking back we see the Sun as a fiery ball surrounded by its planets orbiting in a plane (Fig. 2.2).

The Sun is a star, i.e., it shines from its own power, whereas the planets just reflect the sunlight. Like all stars it is a gigantic fusion reactor producing energy by the fusion of atomic nuclei in its interior, and radiating light and heat away from its hot surface. From a distance of 10,000 billion kilometers (10^{13} km), our solar

G. Börner, *The Wondrous Universe*, Astronomers' Universe,
DOI 10.1007/978-3-642-20104-2_2,
© Springer-Verlag Berlin Heidelberg 2011

Fig. 2.1 The Earth, as seen by the crews of the Apollo flights, floats like a splendid, colorful sphere in space (*courtesy of NASA, Apollo 17*)

Fig. 2.2 This picture shows the Sun, and from left to right the planets. Mercury, Venus, Earth, Mars, Jupiter, Saturn, Uranus, and Pluto in approximately the correct ratio of their sizes (*courtesy of IAU*)

system appears lonesome and somehow lost in the gigantic large-ness of space. The distances have grown so big now that to state them in terms of kilometers would lead to large and impractical numbers. Therefore we choose a new scale, the light-travel time:

We measure distances by the time it takes for light to traverse them. The velocity of light is about 300,000 km s^{-1}. From the Moon to the Earth (384,000 km) the light needs a little more than a second (1.28 s), from the Sun to the Earth about 8 min. Thus we say that the Earth–Sun distance is 8 light-minutes.

We have traveled already 10,000 billion kilometers, i.e., one light-year. Now we approach the nearest neighboring star, and when we continue our flight, we meet stars like our Sun again and again, seemingly without end.

Nothing can move faster than light, but in our imagination we can, of course, exceed that speed limit. After 100,000 light-years we have apparently passed the assembly of stars, because we leave a stellar system behind which resembles a flat disk with a central bulge. It contains about 100 billion stars together with gas and diffusely spread out solid matter called "dust" by astronomers. This is our Milky Way, the "Galaxy."

Extragalactic space seems to be empty, but in the distance we can discern a big stellar system next to us – Andromeda, a galaxy at a distance of two million light-years. It appears similar to our galaxy with spiral-shaped extensions. In our faster-than-light travel we meet such galaxies again and again. They seem to fill all of space. But, when we come to a standstill, we find that all those galaxies rush away from each other in rapid flight, like the fragments of a huge explosion. In addition we observe radiation signals from very large distances, signals from a gradually burning out fireball. We cannot see any further, the origin of the cosmic flight remains hidden to us.

This cosmic system in rapid expansion presents an amazing sight, much more complex than the peaceful, uniform distribution of stars in our neighborhood suggests. How can we obtain an intelligible picture of these conditions in space and time?

Deep in thought we return to the Earth, to the starting point of our voyage, and look at the table at which we sit. A solid piece of furniture, no doubt, supporting us reliably. But the solid, brightly polished surface is only seemingly so. As soon as we begin in our imagination a journey to smaller and smaller separations, the polished surface at first turns out to be a rough landscape of valleys and steeply rising peaks. When we penetrate to dimensions of a hundred millionth of a centimeter, we are surrounded by the electron shells of the atoms, orbiting and oscillating electric charges which are arranged in various regular patterns. The electrons dance around a nucleus which carries practically the whole mass of the atom concentrated in a tiny volume which extends only to one part in a hundred thousand of the scale of the electron shell. (These big numerical factors are written by the physicists in powers of ten: one hundred thousand is 10^5, and one hundred thousandth, or one part in one hundred thousand is 10^{-5}.)

Nothing is there between shell and nucleus – just empty space. The table shows itself as not a very solid object. It is a porous, almost empty thing, but it appears solid to us, because from the point of view of an atomic nucleus we are also a porous structure. We also consist of electrons and atomic nuclei, just like the table. The tiny nuclei of the atoms again are composed of protons and neutrons, the building blocks of matter. At this level all inanimate and living things in the world are equal: an assembly of protons, neutrons, and electrons.

Different forms and shapes are created from these identical small building blocks by arranging them in various ways according to the laws of physics.

To be sure, protons and neutrons are not elementary particles yet. If we imagine a look inside those nuclear particles, we find empty space, and point-like particles, the quarks. Both neutron and proton contain three quarks. The electrons, like the quarks, are indivisible point-like particles. Inside the nuclear particles we cannot look at the world as easily as in our normal human environment, where there are tables, houses, cats, human beings, and much more. The elementary particles do not keep such a well-defined identity, they become much more vague, merge in some sense with the forces acting upon them, and can no longer be discerned as tiny objects in space. It is difficult to write about impressions on the way to even smaller dimensions, because we do not have adequate conceptions for that in the classical world. We can still regard the quarks as a form of material objects, but if we try to probe more deeply, we find that the material properties fade away. The seemingly point-like particles are actually concentrated packets of energy produced by the vibrations of a diminutive string. To be sure, we are also gradually losing any orientation in space and time, as we venture into this regime of dimensions less than 10^{-33} cm. It seems that even space and time perish in the ups and downs of string vibrations.

Deeply impressed by this vision we return to our reliable environment of solid objects.

Does this view of the world on its largest and smallest dimensions truly describe reality? First of all it is nothing but pictures and mathematical constructions invented by us to help us grasp the world around us. Of course, these images are coined by our senses, our reason and mind, i.e., by our brain which has been shaped during a long biological evolution, and is thus dependent on the world too. All scientific insights, and our daily experiences as well, are filtered by these conditions of our sensory equipment. Nevertheless, it looks as if a reality existed independently of ourselves, as if we could succeed in unraveling its properties step by step, even demonstrate and explain its counter-intuitive aspects.

At this point let us leave these preliminary remarks, and turn in detail to some of the things we have seen on our excursion.

2.2 Cosmology

The big-bang model of the universe can be comprehended in illustrative pictures, but the description of the path from astronomical observations and theoretical considerations to that model requires a discussion of many astronomical and physical details. We have to inspect the wealth of detailed results which can be combined to yield the present-day view of the cosmos.

If we want to understand how well the standard model is established, we have to consider stellar evolution, the spectral analysis of the light received from distant galaxies, the properties of the cosmic microwave radiation, and some fundamental features of Albert Einstein's theory of general relativity.

2.2.1 The Darkness of the Night Sky

There are a few easy cosmological observations which neither require expensive telescopes nor satellites. The cheapest entry to cosmology is right above you, when you stand in front of your doorstep at night and look at the stars in the sky. Why is the sky between the stars dark?

If the stars were distributed uniformly in space, were shining forever without change, then there would be no gap between the stars. In every direction you would see a star – some close, some far away. The night sky would be everywhere as bright as the surface of a star. The night sky is dark, however, and therefore this assumption about the stars cannot be correct.

Johannes Kepler in 1610 had already noticed that the darkness of the night sky contradicted some older ideas about the structure of the world, especially the view of Giordano Bruno, who held that the cosmos was infinite and unchanging. Later on Kepler's arguments were repeated several times, for instance in 1823 by the physician and astronomer Heinrich Wilhelm Olbers of Bremen. They are named after him "Olbers' Paradox," although they are not paradox, and were not invented by Olbers. Interestingly enough, even at the beginning of the twentieth century most astronomers believed in a static world. This was the motivation for Albert Einstein to look for a uniform, static cosmological model as a solution of his theory.

Today we know that the world is not static, that all stars came into being a finite time ago, and that they will all perish. Therefore there is only a minimal chance to find a star in any direction, and the night sky appears dark.

Besides the light from the stars we also receive radiation from the hot plasma of the early universe which surrounds us in huge distances like a giant hollow sphere. The cosmic expansion stretches all wavelengths, and shifts this radiation out of the visible range into the microwave region of the electromagnetic spectrum. Clearly, it deserves its name "cosmic microwave background (CMB)," and it surely does not disturb the darkness of the night sky. Thus this everyday, or rather every-night, commonplace fact of the darkness of the night sky tells us that the world is expanding, and that the stars have arisen a finite time ago.

2.2.2 The Life-Cycle of the Stars

In dark nights away from city lights we can see the bright band of the Milky Way stretching across the sky – billions of stars which

Fig. 2.3 Obtained with the Hubble Space Telescope (HST) this picture of a region in the constellation "Eagle" (Aquila) shows structures where new stars are forming. Columns of cold hydrogen gas and dust are illuminated by the UV radiation of the newly formed stars. A color composition is chosen such that an optically appealing impression results (*courtesy of the Space Telescope Science Institute (STScI)*)

send out their energy like the Sun. In reality we can, just using our eyes – the "unarmed" eye as militant astronomers love to put it – see only about a thousand of the nearest stars.

Our Sun is a very typical star. It was born in the condensation of an interstellar gas cloud. The initial clump of gas contracted more and more under the action of its own gravity, until finally the center became hot and dense enough to start the fusion of hydrogen into helium (Fig. 2.3).

In this reaction four atomic nuclei of hydrogen, i.e., four positively charged nuclear particles (four protons), merge to form one atomic nucleus of helium which consists of two protons and two neutrons. The mass of the four individual nuclear particles adds up to a mass larger than that of the helium nucleus by 0.7%. According to Einstein's famous equation ($E = mc^2$: energy is equal to mass times velocity of light squared) this mass deficit is translated into energy by the fusion reaction. The fusion energy per gram of matter is about a million times larger than the energy set free in a chemical reaction like a combustion of fuel or an explosion. Only the explosion of an hydrogen bomb displays the dramatic example of fusion energy suddenly set free.

In the interior of the Sun the hydrogen fusion runs as a slow, controlled process. It has made the Sun shine for 4.5 billion years. In about another 5 billion years the hydrogen at the Sun's core will have been used up. Nuclear fusion in the center will stop when about 12% of the hydrogen supply are consumed. Then hydrogen starts to burn in a spherical shell around the central region. The Sun tries to establish a new equilibrium and blows up its outer layers enormously – it turns into a "red giant." When the Sun reaches this stage, its outer layers will extend to the Earth's orbit. The Sun, like any other similar star, will exist only for a relatively short time, only about 500 million years, in the red giant stage. After that, helium burning starts in the center followed by further short-lived fusion reactions. Eventually the outer shell of about one quarter of the mass is expelled. The remaining core shrinks to a very dense object of about the Earth's dimension – to a "white dwarf." As a white dwarf the Sun will shine with a bluish light scarcely brighter than the full Moon on the burnt-out Earth. This evolutionary history can be accurately predicted, because it simply follows from the laws of physics which govern the nuclear reactions inside the Sun. But there is no reason to panic – the Sun will exist as a quietly shining star for a substantial amount of time. Mankind has just started its evolution as an intelligent life-form. If they do not perish prematurely, our offspring will advance, during the billions of years of their future, far beyond Earth to distant solar systems. Even if now the Earth was the only planet carrying life, there would be sufficient time for life to spread out over all of the Milky Way, and even to other galaxies.

The evolutionary path of other stars can deviate considerably from that of the Sun.

Stars with larger mass produce more energy in their interior, are more luminous, and remain for a much shorter time span in the phase of hydrogen burning. A star which is about 10 times more massive than the Sun enters the red giant stage already after 10 million years. A smaller star with about 10% of the Sun's mass uses its nuclear fuel very economically, shines quite faintly (only at about one thousandth of the solar luminosity), and exists for about 10,000 billion years.

The final stage of massive stars is quite dramatic: Since the core of a massive star contains too much mass for a stable white dwarf, it will collapse further driven by the relentless pull of gravity, and end up at a much smaller radius, and a much higher density. A neutron star (an extreme object resembling a gigantic, solar mass atomic nucleus with a radius of 10 km) or a black hole may be the final state. In a certain mass range the core may even be disrupted completely. In any case a huge eruption is triggered, the so-called supernova explosion, which hurls the outer parts of the star into the surrounding space. A supernova shines for some time with extreme brightness, often surpassing the whole host galaxy in luminosity. The remnants in many cases radiate actively as pulsars or X-ray sources (Fig. 2.4).

Our Milky Way is populated by all these different types of stars: Blue, very bright, massive, and short-lived stars which are formed again and again from gas and dust, many stars like our Sun, red stars of small mass which are long-lived and faint, luminous red giants, white dwarfs, pulsars, and X-ray sources, all are contained in this huge stellar system.

Fig. 2.4 The Crab nebula shown on this HST image is the remnant of a supernova observed in 1054 AD by Chinese astronomers. A "pulsar" at the center of this nebula emits periodic radio signals with a period of 33 ms. This pulsar is a neutron star which rotates 30 times per second around its axis (*courtesy of the STScI*)

Every 100 years or so a supernova explodes in a typical galaxy. The last time such an event could be seen in our Milky Way was in 1604 (Kepler's supernova). In the year 1987 a spectacular supernova could be observed in the Large Magellanic Cloud, a small stellar system close to the Milky Way at a distance of only about 180,000 light-years. For astronomers supernovae are naturally of great interest, but they are important for the whole of mankind too:

The heavy elements, present also in the human body, like carbon, oxygen, silicium, iron, etc., have all been formed in the interiors of massive stars, and have been distributed in space during the explosion of these stars. In that sense we are children of the supernovae. Only the lightest elements hydrogen and helium were created in the early universe, all heavier elements were brewed in stars.

2.2.3 The Galaxies

We can recognize many more stars with a telescope, and with it we also see that besides the Milky Way there are many fuzzy luminous spots in the sky, which turn out to be stellar systems like our Milky Way, "galaxies" as they are named.

Galaxies appear in a great variety of shapes (Figs. 2.5– 2.8):

Systems with spiral arms of similar size as our own galaxy (such as M31, the galaxy named "Andromeda" at a distance of two million light-years) or elliptical galaxies without spiral arms resembling an elliptical or nearly circular small disc are frequent. Elliptical galaxies can be very massive (some contain 10^{13} solar masses, a hundred times the mass of the Milky Way), but there exist also very small dwarf galaxies of similar appearance, but with a mass of only a few million solar masses. Light needs about 100,000 years to cross the Milky Way or the Andromeda galaxy, and the separation between galaxies is typically ten times larger, about a million light-years.

The huge distances between galaxies have an interesting consequence: The farther away a galaxy is, the earlier in its history we can see it. When we observe Andromeda, we do not see what happens there right now but we see what has happened 2 million years ago. It is thus impossible for us to observe the universe

Fig. 2.5 The Andromeda galaxy is the nearest large spiral galaxy, a neighbor of the Milky Way at a distance of 2 million light-years. Viewed from far away our own Milky Way would probably look quite similar. This image obtained from data of NASA's satellite WISE (Wide field Infrared Survey Explorer) shows Andromeda in infrared light at different wavelengths (from 4 μm (*blue*) to 22 μm (*red*)). Mature stars show up in *blue, yellow,* and *red* colors indicate regions where dust is heated by newborn, massive stars (*courtesy of NASA*)

Fig. 2.6 The spiral galaxy NGC4622 rotates clockwise – from the image one might draw the opposite conclusion (*courtesy of STScI*)

Fig. 2.7 The galaxy cluster A2218 is a gravitational lens. It deforms the images of distant galaxies to elliptical shapes (*courtesy of STScI*)

Fig. 2.8 This image of a small area of the sky (about 1 arcmin × 1 arcmin wide), the so-called Hubble Deep Field, is the deepest look into the universe in optical light reached so far. The picture does not only belong to the treasures of astronomy, but it is also a treasure of mankind. The small area in the constellation Ursa Major contains about 2,500 galaxies of all types – disclike elliptical, spiral, and irregular galaxies. The HST was pointed at this location for 10 consecutive days (*courtesy of STScI*)

in its present state. Astronomers investigate the world similar to archaeologists, who dig into deeper and deeper layers as well as earlier and earlier times. The advantage is that one can look directly at the evolution in time, the disadvantage is, of course, that one can never see the whole at one particular instant of time.

These aspects are due to the finite velocity of light, and they remain valid for small distances. But their consequences in that case are insignificant: Light takes 8 min from the Sun to the Earth, but in 8 min nothing much changes in the solar system even though the Bavarian Broadcasting Corporation (BBC) claims that "in 15 minutes the world may change."

Figure 2.8 shows a picture of many galaxies taken by the Hubble Space Telescope (HST), about 2,500 in a narrow region of the sky. Extrapolating to the whole sky astronomers can safely estimate that the volume of space accessible to observation contains about ten billion galaxies. Each individual galaxy with its billions of stars is in itself an interesting system, but in cosmology it is regarded as a test particle useful for exhibiting some, perhaps really existing, global properties.

2.2.4 The Expansion of the World, and the Cosmic Microwave Background

The modern view of the cosmos derives from insights into a fundamental property of the galaxies. In the 1920s the American astronomer Edwin P. Hubble found that the spectral lines of atoms measured in galaxies do not coincide with those measured in laboratories on Earth. Instead almost all galaxies (with the exception of Andromeda, and some small companions of the Milky Way) exhibit a shift of their spectral lines toward longer wavelengths (toward the red end of the spectrum) by a factor $(1 + z)$. For each galaxy this factor is a characteristic quantity, all its spectral lines are stretched in wavelength by the same factor. z itself is called the "redshift" of the galaxy.

The redshift z increases with the distance of the galaxy.

The Doppler effect explains this phenomenon quite naturally: For a source of light moving away from us, the wavelength of the signal received is longer than the emitted one. We may conclude that almost all galaxies are moving away from us, the farther away they are the faster they move. In fact, the velocity is proportional to the distance according to the famous relation discovered by Edwin Hubble

$$cz = v = H_0 d.$$

The quantity H_0 was named "Hubble Constant" to honor Edwin Hubble's discovery.

In this equation which describes the increase of the velocity $v = cz$ with distance d, we also find c – the velocity of light.

To measure the cosmic expansion, i.e., the flight of the galaxies, you need to measure the distance d and the redshift z of only one galaxy, at least in principle. In practice you meet a few difficulties: The astronomers know precise distances only to relatively close galaxies, and those have proper motions – induced by local mass concentrations – superimposed on the cosmic expansion motion.

Andromeda's proper motion even dominates over its cosmic motion. It approaches the Milky Way, and its spectral lines are therefore blue-shifted.

Hubble, and many astronomers after him, have used pulsating stars for cosmic distance measurements. "Cepheids" (named after the star δ-Cephei) change their brightness rhythmically; they pulsate with periods of hours to days. Slow pulsation signifies high luminosity, and two stars with the same pulsation period have the same luminosity. Therefore the measurement of the pulsation period and the brightness of a Cepheid is sufficient to determine its distance, if a few stars are known with precisely determined distances to calibrate the relation between pulsation period and luminosity. Cepheids can provide very precise distance determinations, but unfortunately for many years Cepheid stars could not be measured at the cosmic distances, where the expansion velocity dominates over local, peculiar velocities.

It was expected that the situation would improve decisively, when new telescopes, especially the space telescope "Hubble," would extend the classical Cepheid method out to a distance of 20 Mpc (the unit Mpc – "Megaparsec" – is about 3.26 million light-years). This is the distance to the center of the Virgo cluster of galaxies. At the edge of this huge system of thousands of galaxies lies the Milky Way. Unfortunately the Virgo cluster has shown itself as a relatively complex structure, where the center of mass is difficult to determine. Thus a spread-out range of values for the Hubble constant had to be accepted, namely

$$H_0 = 80 \pm 22,$$

in units of velocity (given in kilometers per second) per megaparsec. These units are favored by astronomers, and they have an easy interpretation: At 1 Mpc distance a galaxy recedes with a velocity of 80 km/s.

In view of the uncertainties involved in the Virgo distance, a new method has been developed which allows us to reach out to far greater distances without intermediate steps.

This approach makes use of the high luminosity of certain types of stellar explosions, the supernovae of type Ia (SNIa). Their spectra do not contain lines of hydrogen, only higher elements such as helium or carbon are present. Stars which end their existence as SNIa evidently have gone through a long time of evolution. They have burnt their hydrogen supply and the stellar material is essentially carbon and oxygen. Very probably these are white dwarfs, compact stars with a radius like the Earth and a mass like the Sun. The luminosity of such a supernova increases rapidly, reaches a maximum within a few days, and then decreases.

The explosion produces radioactive nickel (^{56}Ni) which decays via cobalt (^{56}Co) to iron (^{56}Fe), and thereby supplies the energy of the luminous phenomenon. According to theory the optical luminosity of a SNIa is due to the thermalization of high-energy gamma rays produced during the decay of nickel and cobalt.

SNIa are very bright. They can be observed at great distances far beyond Virgo. In addition they are good distance indicators, although they do not all have the same peak luminosity. One must expect some variation, because the luminosity depends on the amount of nickel produced, and this can vary according to the conditions in the star when it explodes. There is, however, a very helpful property: The observers found that there is a strong correlation between the maximum luminosity and the shape of the supernova light curve, especially the decline of the brightness. Rapidly decaying light curves belong to less luminous supernovae, while slowly decaying ones are more luminous at maximum. This empirical relation can be quantitatively fixed, and thus the maximum luminosity can be calibrated making supernovae Ia a precise indicator for cosmic distances.

During the past 12 years astronomers have succeeded to detect very distant type Ia supernovae systematically, and to measure the rapid increase as well as the decline after maximum of their brightness.

The collaboration of many observing stations around the world had to be organized such that each supernova could be traced by a big telescope immediately after its detection. Two large groups of observers, the "High-z Supernova Search Team" and the "Supernova Cosmology Project," have independently pioneered this research.

Figure 2.9 displays a Hubble diagram for a large number of SNIa. The data points below a redshift of $z = 0.1$ agree very well with the linear-Hubble relation. This leads to a determination of the Hubble constant

$$H_0 = 70 \pm 10.$$

The Hubble constant given in these astronomical units defines a characteristic time by its inverse $1/H_0$. This "expansion time" amounts to about 14 billion years with an observational uncertainty of about 10%. The expansion of the system of galaxies which we observe today has started 14 billion years ago provided the galaxies have moved with constant velocity. At that time all the galaxies we can see now must have been very close together.

The measurements of the cosmic expansion gain special importance, if we take another cosmological discovery into account:

Two scientists, Arno Penzias and Robert Wilson, working at the Bell laboratories discovered rather accidentally in the year 1964 a radiation signal while they were calibrating a special antenna for microwave transmissions. The radiation at a wavelength of 7.15 cm was apparently of cosmic origin, because typical temporal variations as they are shown by individual sources were absent. Further measurements gave evidence that the radiation with wavelength between 1 mm and 10 cm arrives from all directions with nearly equal intensity, and that its spectral distribution

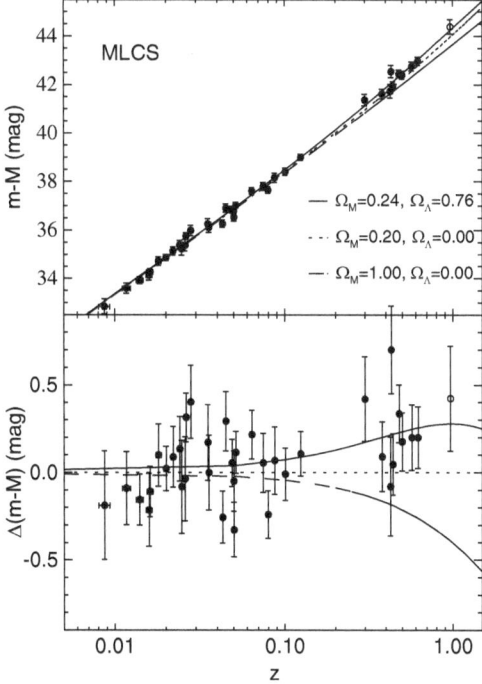

Fig. 2.9 The Hubble diagram for supernovae Ia shows the data points as they depend on distance and redshift z. Along the vertical axis the distances are given in a logarithmic unit loved by astronomers, the so-called distance modulus. The redshift z is displayed on a logarithmic scale along the horizontal axis. In the *top panel* you can see, how well the Hubble relation holds for SNIa at small redshifts (for z less than 0.1), while at large redshifts deviations from the linear Hubble relation occur. Supernovae at $z \approx 1$ clearly lie above the straight line of the linear relation indicating that they are more distant than their redshift would tell. The astronomers regard this as evidence for an accelerating cosmic expansion. Cosmological models with a positive cosmological constant have such a property. The graphs for three different cosmological models in the *upper panel* differ significantly at large z. The *lower panel* displays these differences referred to the model without a cosmological constant. At high redshift a model with a positive cosmological constant gives the best fit (after Riess et al., 1998, Astrophys. J. **504**, 935)

follows the law found by Max Planck around 1900 for the radiation emitted by a body in thermal equilibrium with its surroundings. Penzias and Wilson have received the Nobel prize for physics a few years later, since it immediately became clear that their discovery had a great impact on our knowledge of the cosmos.

Since it obeys Planck's formula, the cosmic microwave background (CMB) can be characterized simply by a temperature. The satellite COBE (Cosmic Background Explorer) has yielded measurements of the CMB spectrum over 2 years which determine this temperature precisely as:

$$T = 2.728 \pm 0.002 Kelvin.$$

(Kelvin is a temperature scale like degrees Celsius, shifted such that zero Kelvin corresponds to the absolute zero of −273.2 degrees Celsius.)

Within the measurement errors no deviations from an ideal Planckian spectrum could be found. Thus, the CMB defines the present temperature of the universe (see Fig. 2.10).

CMB and Hubble expansion taken together point at an interesting aspect of the history of the universe: If the galaxies now

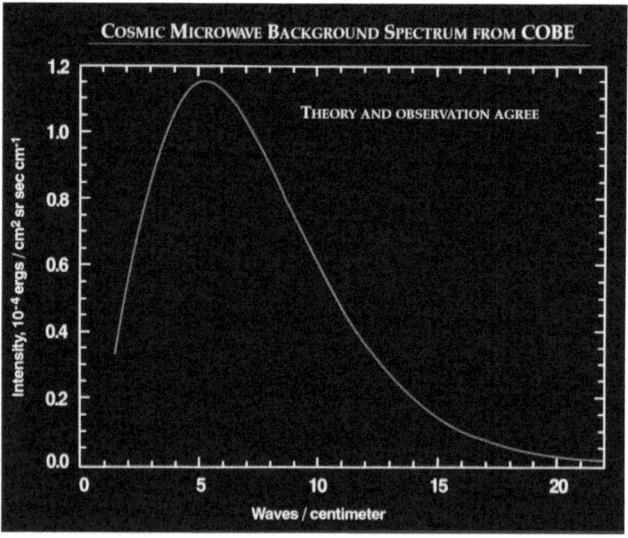

Fig. 2.10 The spectrum of the cosmic microwave radiation (CMB) as it has been registered by the satellite COBE fits perfectly the formula for thermal radiation with a temperature of 2.728 K, i.e., about 2.7 degrees above the absolute zero point of temperature. Measurement uncertainties are less than 2 mK (±0.002 K). This radiation is a natural consequence of the "hot big bang" model: It is the relic radiation of an early phase, where an almost uniform hot plasma was in thermal equilibrium with the radiation field. This cooled down because of the cosmic expansion (with permission of the COBE collaboration; Mather et al., 1990, Astrophys. J. **354**, L37; Fixsen et al., 1996, Astrophys. J. **437**, 576)

fly away from each other, they must have been closer together at earlier times. Then also the radiation must have been denser, more compressed, and hotter in the past. The conclusion seems inescapable that there has been a hot and dense early state of the universe. In the hot early universe galaxies and stars could not survive, and all that existed was a hot and dense mixture of matter and radiation.

The expansion time of 14 billion years derived from the Hubble diagram of type Ia supernovae defines the time in our past, when galaxies appeared out of the "primeval soup," and began their flight in space.

Even if this interpretation of the CMB and the general expansion sounds very plausible, we must be aware of the fact that it is not just a consequence of the observations. Theoretical conceptions are inextricably mixed into it. The universe as a whole is actually a theoretical construction, and a very special object of research, unique and unreproducible. Every physicist would be unhappy if he had to build his theories on a single experiment which could not be repeated.

But the situation is even more difficult, because we, the observers, are part of this object "universe," and living inside it we can only perceive a section limited in space and time. We assume that the part we can observe is typical for the whole – if that exists at all. But this is by no means sure. In cosmology we must work in the context of a given theory, and try to sketch a model of the cosmos using this theory, and the observations and measurement results. Only with the help of the model observations can be interpreted, and new observations and tests of the model can be suggested.

2.2.5 The Cosmological Model

The search for a simple model of the flight of the galaxies will focus on an easy mathematical representation of the uniform expansion. It seems reasonable to avoid the point of view which would put us into the center of the universe with all the galaxies moving away from us. There is no compelling cause for that, and therefore a better description would be to assume that the cosmic expansion looked the same observed from any galaxy, similar to what terrestrial astronomers observe.

Fortunately the uniform spreading out of the celestial bodies can be modeled by simple solutions of Einstein's theory of gravitation, the "theory of general relativity" (GR). Within the models the distribution of matter is taken into account only approximately, as an average matter density, and not exactly as a large system of galaxies and stars.

Let Albert Einstein himself comment on that: "The metric character (the curvature) of the four-dimensional space–time continuum is determined according to General Relativity at each point by the matter and its state in that point. The metric structure of the continuum must therefore be extremely tangled up due to the non-uniformity of the matter distribution. But if we care only about the structure on large scales, we may imagine the matter uniformly distributed over huge volumes, such that the distribution of the density becomes an enormously slowly changing function. We thus proceed similar to geographers, who approximate the Earth's surface which in small details is shaped extremely complex by an ellipsoid."

("Der metrische Charakter (Krümmung) des vierdimensionalen raumzeitlichen Kontinuums wird nach der allgemeinen Relativitätstheorie in jedem Punkt durch die daselbst befindliche Materie und deren Zustand bestimmt. Die metrische Struktur dieses Kontinuums muss daher wegen der Ungleichmäßigkeit der Verteilung der Materie notwendig eine äußerst verwickelte sein. Wenn es uns aber nur auf die Struktur im Großen ankommt, dürfen wir uns die Materie als über ungeheure Räume gleichmäßig ausgebreitet vorstellen, so dass die Verteilungsdichte eine ungeheuer langsam veränderliche Funktion wird. Wir gehen damit ähnlich vor wie etwa die Geographen, welche die im Kleinen äußerst kompliziert gestaltete Erdoberfläche durch ein Ellipsoid approximieren.")

It is favorable for model-building that the cosmic expansion does not depend on the way matter is distributed in a certain volume of space. Quite inhomogeneously condensed in galaxies and stars or uniformly spread out – it makes no difference; only the mean density, i.e., the mean mass per unit volume, counts. Therefore we can in a first approximation neglect all the structures, and regard the total mass in a certain volume of space, as e.g., in a giant sphere enclosing many galaxies, as a tenuous gas.

In fact, this gas has such a tiny density that it almost represents an ideal vacuum – only about one atom is contained within $1\,m^3$. Cosmologists find such a small density, when they add up the masses of the galaxies. In addition, there is strong evidence, as we shall see later, of the existence of nonluminous, so-called dark matter, and of a quite different component named dark energy. All these various types of matter and energy form a "cosmic substrate," as we might call it. This appears in the cosmological models only as a uniform density, i.e., as matter or energy averaged over large volumes of space. That is an approximation, but a very good one, as many computations have demonstrated.

A convenient simplification can be introduced, as is commonly done, by characterizing the various density components by non-dimensional numbers, i.e., by their ratio to a reference density which can be constructed from the gravitational constant G, and the Hubble constant H_0. Both quantities can be combined such that a term with the dimension of a mass density (grams per cubic centimeter) results:

$$\rho_c \equiv \frac{3H_0^2}{8\pi G}.$$

This reference density ρ_c is often called "critical density." Inserting the measured value of H_0 one finds that this critical density corresponds to a matter content of about ten hydrogen atoms per cubic meter. This is an excellent "vacuum" not yet achieved so far in terrestrial laboratories.

Following these approximations cosmologists use simple cosmological models, so-called Friedmann–Lemaître models (FL models for short; named after Alexander Friedmann (1922) and Georges Lemaître (1927), who were the first to derive and interpret these special solutions of Einstein's theory of gravitation): The expansion is thought of as the spreading out flow of an idealized uniform matter, comparable to a fluid with homogeneous density $\rho(t)$ and pressure $p(t)$ which change with time. The fluid particles can be imagined as representations of the galaxies in this picture.

Their separation increases with time as they follow the general flow pattern in the expanding cosmic material. This expansion can continue without end, or it can reach a maximum and then turn into a contraction (see Fig. 2.11). The difference in

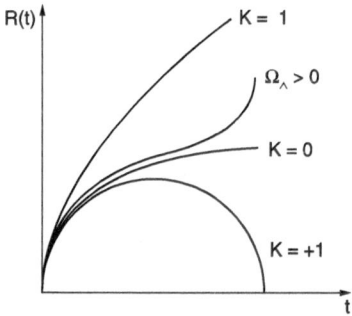

Fig. 2.11 In simple cosmological models, the Friedmann–Lemaître models, the separation of two particles of the cosmic medium changes proportional to a function of the time $R(t)$ in the way shown schematically in this figure. The number K characterizes the curvature of space ($K = +1$: spherical; $K = 0$: Euclidean; $K = -1$: hyperbolic). The curve labelled $\Omega_\Lambda > 0$ describes a model with a positive cosmological constant (see text) which seems to fit the observations very well. All models have the property that there are only changes with time. There are no variations in space. In all models there occurs a zero point of time, where all distances between objects go to zero, density and temperature become infinite. This singular point therefore lies outside the range of validity of the models depicted here

behavior is caused by the amount of matter, radiation, and other possible forms of energy in the cosmos.

The cosmic density ρ_0 is generally replaced by its ratio to the critical density, and thus written as a pure number, the "density parameter"

$$\Omega_0 \equiv \frac{8\pi G}{3H_0^2}\rho_0 \equiv \frac{\rho_0}{\rho_c}.$$

When the density is given in this way, as a dimensionless number, the fact is nicely illustrated that in these cosmological models there is no other dimension than the Hubble constant.

Not only the massive objects contribute to the total density, but any other form of energy. All the different components can be added up to a total density parameter Ω which is the sum of individual contributions each given as a fraction of the critical density.

If Ω is less than 1, i.e., if the density is below the critical one, the expansion will go on forever, but for Ω greater than 1 the expansion may turn over into a contraction, leading to the collapse of everything into a final singularity, a big "crunch."

These possibilities can also be seen in the graphs of Fig. 2.11. Which case corresponds to the real universe? Astronomers try to find out by measuring the cosmic density.

2.2.6 Accelerated Expansion

The supernovae plotted in Fig. 2.9 seem to be more distant at large redshifts than would correspond to the linear Hubble relation. Apparently the distance between us and these supernovae has grown faster than it would have, if these objects had moved with constant velocity. The expansion of the cosmos is accelerating, whereas in fact a slowly decelerating motion might be expected, if all the moving galaxies attracted each other gravitationally.

This accelerated expansion might be caused by a constant, positive energy density which would act like a repulsive gravitational force on cosmic scales. There is nothing new to a quantity of this kind. Albert Einstein already had introduced it in the equations of his theory of GR with the aim of deriving a world model for a uniform and infinite distribution of stars. Such an infinite, static system was the general view of the cosmos around 1915. Einstein defined a "cosmological constant Λ," a quantity which at present is generally written as Ω_Λ, a cosmological constant density parameter ("ccd" for short), where

$$\Omega_\Lambda \equiv \frac{\Lambda}{3H_0}.$$

As we have said already, a positive cosmological constant acts like a repulsive force which may, if it has the right magnitude, completely balance the attractive force of gravity.

When Edwin Hubble discovered the expansion of the universe, and when Alexander Friedmann showed that GR has solutions corresponding to expanding cosmological models, Einstein wanted to erase the cosmological constant from his theory. He felt sorry for having introduced it, his "biggest folly" ("größte Eselei" in German) as he said. But now this quantity has been finally established again due to the astronomical measurements of the Hubble expansion, albeit with a smaller value than the one postulated by Einstein. The equations of GR demonstrate that Ω_Λ accelerates the expansion, if it is bigger than half the matter density $(\Omega_m/2)$.

The best fit to the data in Fig. 2.9 is achieved, if one chooses values of $\Omega_m = 0.3$ and $\Omega_\Lambda = 0.7$ for the cosmological model which clearly meets the conditions for accelerated expansion.

We will bring up further evidence for a positive cosmological constant, when we discuss the anisotropies of the CMB. In spite of the impressive observational indications, theoreticians feel somewhat uneasy about the existence of a cosmological constant. This component of the cosmic substrate is not really some "stuff" filling space like a gas, it is rather a property of empty space, a kind of inner tension which is relaxed and balanced by the expansion of space. Later on we will consider in detail the attempts to explain this mysterious quantity, especially the interpretation favored at present as the energy density of a field. Anyway, the name "dark energy" appears well chosen, since it hints at hidden action without accompanying luminous phenomena and at the darkness surrounding the true nature of this quantity.

2.2.7 Curved Space

In Friedmann–Lemaître models there are three different theoretically possible types of curved space: At any fixed time three-dimensional space is either the space well known from everyday experience, flat with Euclidean geometry, or a space with constant positive curvature, or a space with constant negative curvature. The conception of "curved spaces" is difficult, and without a recourse to mathematical expressions not easily understood. We might try to obtain a picture of those spaces in our imagination, if we think about the two-dimensional counterparts reducing the real spaces by one dimension. The three different types of space correspond then to the geometrical picture of a plane (this is the Euclidean space with curvature zero), the surface of a sphere (positive curvature), or a saddle-like surface (negative curvature) (Fig. 2.12).

Spherical and saddle-like space are surely more difficult to imagine, than the flat, infinite space. The spherically curved space is closed like the surface of a sphere: One returns to the starting point, if one continues to go straight ahead. "Straight ahead" means, of course, to follow a great circle (i.e., a circle with its center at the sphere's center) on the surface of the sphere.

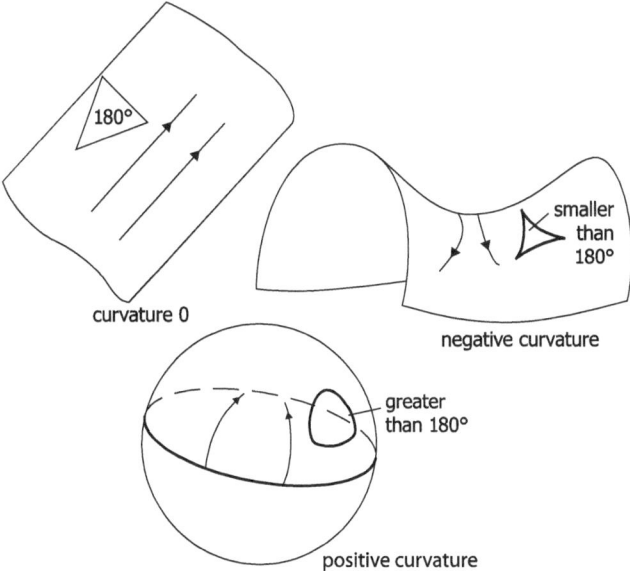

curvature 0

negative curvature

positive curvature

Fig. 2.12 Curved spaces can be illustrated as surfaces in 3-space, if one dimension is cut out. Three types of surfaces with constant curvature can be discerned: The plane corresponds to Euclidean space with curvature zero, where the sum of interior angles in a triangle is 180°, and where parallel lines never intersect. The spherical surface gives a picture of a space with positive curvature, where the sum of the angles in a triangle is greater than 180°, and where "parallel" lines meet the poles. Similar to the saddle-like surface is a space of negative curvature, where parallel lines diverge and the sum of angles in a triangle is less than 180°. The type of 3-space and the way the expansion develops are closely connected according to Einstein's theory

One never meets a boundary, because no boundaries exist on the surface of the sphere. The two-dimensional analogy is unfortunately somewhat unconvincing, because we must completely forget about the space outside of the surface of the sphere – only the surface itself exists and forms all of space. For three-dimensional space we have to imagine a spherical surface in four-dimensional space – not easy, even after long training.

The total volume of a spherical space is finite, just as the surface of a sphere has a definite, finite area. Flat spaces and saddle-like spaces are infinite and open, and by going straight you will never return to the starting point. In FL models it is the matter and energy density which curves space. Larger density leads to larger curvature, e.g., to a smaller sphere in the case of positive curvature.

The idea that space may be curved is a basic aspect of Albert Einstein's theory of GR: Space and time are not fixed and absolute, but defined by the masses and energies present. A massive body distorts the space–time measure in its environment, that is it influences the way clocks run and it changes the measuring rods. Vice versa the space–time geometry acts on the dynamics of the bodies. The interaction between all the masses and energies finally results in the cosmological model.

From this point of view it is absolutely astonishing that the interaction of all things in the cosmos leads to the smooth geometry of a space of constant curvature, or even of a Euclidean space.

An intuitive picture of the expansion might be given by imagining the stretching of an elastic plane, spherical, or saddle-like surface. Let us look for example at the spherical surface: The expansion is illustrated as a uniform inflation of the closed, finite surface, similar to the puffing up of a rubber balloon. "Galaxies" can be represented by marking points on the balloon. When the balloon inflates, the marked points move away from each other. The distances between points grow with the inflating balloon, although their positions (longitude and latitude) on the spherical surface remain the same. The distances change because the elastic material is stretched. This appears to be quite a useful intuitive illustration of the conditions as they are described by Einstein's theory: Distances grow because the space–time structure changes, not because the galaxies themselves move. Thus for an imagined two-dimensional observer in one of the "galaxies" on the balloon surface, the impression arises that all the other galaxies move away from him.

You can imagine such an observer on any "galaxy," and from any point of the rubber balloon he will obtain the same view of the expansion, if the points are distributed homogeneously on the surface.

For galaxies close to us the linear Hubble law holds, while for distant galaxies the curvature of the space–time becomes significant. The redshift can no longer be explained by just the Doppler effect of galaxies moving away from us, but in reality the properties of light propagation in FL models must be taken fully into account: Changes of distances by the cosmic expansion are proportional to a function of time $R(t)$ which is in our intuitive picture just the radius of the balloon.

Light propagates in the space–time geometry in a way such that one plus the redshift z is equal to the ratio of the radius $R(t_0)$ at the present time t_0 to the radius $R(t_e)$ at the time t_e of emission of the signal. (In mathematical terms $1 + z = R(t_0)/R(t_e)$; for times t_e, close to the present time t_0, i.e., for close galaxies, the Hubble relation can be derived from this expression with $H_0 \equiv (dR(t_0)/dt)$ at t_0.)

If we look to the past, we see the balloon shrink. Toward the big bang all the points marked on the surface move closer and closer to each other. On the surface which represents our world there is no special point which marks the location of the beginning of the expansion, of the big bang. All points on the surface are always there, even arbitrarily close to the big bang and even on an arbitrarily small balloon. In the intuitive two-dimensional model one might think that the center of the spherical balloon is the point, where the big bang happened, but this point outside of the two-dimensional surface of the balloon does not belong to our two-dimensional world.

Moving back in time toward the big bang any finite separation of two particles goes to zero. As the particles pile up more and more, density and pressure grow beyond any limit, and become infinite at the initial state which is generally designated as the "big bang." Even theoretically we cannot follow the run of events further into the past, because the conceptions of the theory, even of time and space lose their meaning. This initial "singularity" marks the beginning of the world: Everything we observe now has come into being in a primeval explosion about 14 billion years ago. In the beginning there was infinite density, infinite temperature, and an infinite rate of expansion!

2.2.8 Redshift and Evolution in Time

The situation of the astronomers in a world described by an FL model is as follows: Light signals from distant galaxies arriving here and now have been sent by the source a long time ago. The galaxies are not observed in their present state, but as they were in a previous epoch. The astronomical observations yield a cross section through the history of the cosmos, and its present status can only be derived in connection with an appropriate model.

In our two-dimensional balloon illustration we can mark the observable region by a circle around our position. Objects within the circle can be observed, because light signals emitted by them can be received by us. The circle designates our "horizon," beyond are regions inaccessible to our observations. But our horizon grows with the velocity of light, its radius proportional to time, because light signals traveling with the velocity of light can reach us from more and more distant territories. On the other hand the balloon itself inflates – this expansion depends on the matter and energy densities. As long as matter and radiation are the dominant components of the cosmic substrate, the balloon grows more slowly than the circle representing the horizon, and new areas continuously come into the horizon. For matter the distance between two particles changes with time t as the power $t^{2/3}$, for radiation as the square root $t^{1/2}$, while the size of the horizon grows as t.

If the expansion is dominated by a cosmological constant, the rubber balloon stretches more rapidly than the horizon grows, and gradually individual galaxies disappear from our field of vision. Correspondingly our vision loses in range, when we follow the expansion back into the past. In the cosmos dominated by matter and radiation the horizon shrinks much faster than the universe contracts. This leads to the curious conclusion that as we approach the big bang there is less and less of the world within our horizon.

The redshift of the light of a distant galaxy is a direct measure of the cosmic expansion, because the universe has grown by the factor $(1 + z)$, since the time, when light has been emitted by the galaxy with redshift z.

Observations of galaxies with a redshift of $z = 6$ tell us that the universe had one seventh of its present size when that light had been emitted. The CMB tells us of an epoch with a redshift of 1100. The cosmos now is 1100 times as big as it was then. Clearly this implies that matter and radiation were much denser when the CMB originated than today.

2.2.9 A Time-Lapse Picture

Let us compress the history of the universe into 1 year. Each month then equals a bit more than a billion years in reality. Let us imagine that as the bells are ringing to welcome the new year our world starts with a big bang. The primeval substance, a radiation filling all of space homogeneously with enormous density and temperature, was without structure, but by the momentum of the mysterious initial explosion it expanded and cooled. Already in a tiny fraction of the first second of the 1st of January, matter was created: Elementary particles and soon after the simplest atomic nuclei, hydrogen and helium, were formed. Before the end of January radiation and matter decoupled and the galaxies formed. The first generations of stars in the galaxies brewed the higher chemical elements in their interiors, and hurled them – partly in the form of dust – during the final supernova explosion into the surrounding gas. Carbon was formed most abundantly; this was the basis for the formation of complex organic molecules on dust grains in the vicinity of stars.

In the middle of August our solar system formed out of a collapsing cloud of cool gas and dust. A day later the Sun was more or less in its present state supplying the planets with a pretty steady flow of radiation from its hot surface of 6,000 degrees. The hot solar radiation could be radiated away at a much lower temperature by the Earth, since the interstellar and interplanetary sky was dark and cold. These conditions on Earth permitted the build-up of complex chemical, and then biological structures. The middle of September saw the formation of the first solid rocks on the Earth's surface, and in those oldest rocks we find nowadays first traces of life: fossil one-cell organisms. Already in early October fossil algae developed, and in the course of the next 2 months a huge variety of plants and animals arose, at first in water. The first vertebrate fossils date from the 16th of December. On the 19th of December plants settled on land. On December 20 the landmasses of the continents were covered with forests. Life generated an oxygen-rich atmosphere for itself which shielded it from ultraviolet light, and thus created favorable conditions for even more complex and sensitive forms of life. Eventually, on December 22 and 23, fish evolved into amphibians which could live on dry land. On December 25 the first mammals arrived. The Alps started their folding up during the night before December 30. During the night before December 31 the human primates originated from the branch which also carried a twig leading to the present apes. Human evolution carried on with about 20 generations per second. Five minutes before midnight Neanderthal man lived on the Earth, 15 seconds before 12 o'clock Jesus Christ was born, half a second before the first sound of the bell the age of technology began. Here comes the New Year: How will the story continue?

2.3 Formation of Structures in the Universe

2.3.1 Deuterium, Helium, and Lithium

Within the first second after the big bang protons and neutrons formed out of the cosmic primeval soup. From these basic building blocks a chain of nuclear reactions led after further cooling to the atomic nuclei of the light elements deuterium, helium, and lithium. The nucleus of deuterium consists of one proton and one neutron. Below temperatures of 800 million degrees they are bound in a stable configuration. So the temperature of the cosmic structure must have decreased below that threshold – this can be computed to happen after about 3 min – before deuterium could exist as a stable nucleus. Then further protons and neutrons attached themselves to it, and built the nuclei of helium and in smaller number lithium. This attachment of protons and neutrons does not proceed further, because atomic nuclei with five or eight nucleons (protons and neutrons) are unstable. Therefore heavier elements like carbon or oxygen with 12 or 16 nucleons respectively could not build up. All these elements are produced in massive stars at a later stage in the cosmic evolution.

The big-bang model predicts that the atoms of helium and hydrogen ought to be present with a ratio of their numbers of 1–13, and this agrees well with astronomical observations. Additional assumptions are not necessary to obtain this result. It is a natural consequence of the simple hot big bang, i.e., the FL models.

We may even venture to state that during its first few seconds the universe follows especially well the rules of an FL model. Any small deviation from the expansion law of such a model would lead to a change in the production of helium. The precise measurements presently available to the abundance of helium exclude any significant effect of this kind.

The explanation of the synthesis of helium and deuterium is a big success of the standard big-bang scenario. It is also of great importance, because the production of these elements in stars is not enough: The helium abundance generated by stars is too small in comparison to the measured value of 24% and deuterium is not made in stars at all.

2.3.2 Structure Formation

The explanation of galaxy formation is more difficult, because an obvious discrepancy exists between the uniform, homogeneous cosmological models, and the astronomical observations showing the luminous matter to be arranged in discrete building blocks, the galaxies. Galaxy formation is, in fact, still in many details not understood. This is at present the most active field of research in cosmology.

One basic assumption is to consider galaxy formation as an evolutionary process which leads from initially very small fluctuations of the matter and radiation densities to the structures observed today. Small deviations from uniformity must have existed in the cosmos from the beginning, because nothing complex could evolve from a purely symmetric state.

During this process the initially small inhomogeneities in the cosmic primeval soup are intensified due to their own gravity. Eventually they separate from the general expansion and collapse to dense clumps which follow the expansion as whole objects. This appealing idea meets the following difficulty: Only after the decoupling of radiation and matter, about 400,000 years after the big bang, was it possible for small density contrasts to increase. At earlier times the condensation of matter was prevented by the radiation pressure on the free electrons. When the electrons combined with the atomic nuclei to form hydrogen and helium atoms, the radiation could propagate freely, and the matter could follow its tendency to collapse. The temperature at that epoch was about 3,000 K.

At this time, however, the density contrast of the inhomogeneities, i.e., the ratio of the overdensity of a region to the mean cosmic density, was very small, comparable to the relative amplitude of the fluctuations in the microwave background of about one hundred thousandth (10^{-5}). The density contrast of the matter can grow only by a factor thousand up to now, because the amplitudes increase proportional to the redshift. Thus they could reach only values of a few percent, but not the values characteristic for the density contrast of real galaxies. The conclusion would be that the universe had remained quite homogeneous, that galaxies and stars would not exist. This dilemma motivated cosmologists to investigate nonbaryonic dark matter as a way out for the following reasons: A background of particles of nonbaryonic dark matter does not interact directly with radiation, and is therefore not subject to the strict limit by the CMB anisotropies. Therefore the initial density fluctuations can be bigger than those in normal matter, and they can grow over a larger time span. Finally the dark matter particles would form mass concentrations which attracted and collected the normal matter. The luminous matter, that is to say the galaxies, was like the tip of an iceberg of dark matter which could not be seen itself, but which would determine by its gravity the distribution and velocities of the galaxies.

There is more in these considerations than a well thought-out scheme, because the astronomical evidence for the existence of dark matter is very strong. I will briefly describe some of it in the following.

2.3.3 The Luminous Matter

Visible light is emitted by stars. In the Milky Way and in a few neighboring galaxies stars can be discerned as single objects, more distant galaxies appear as a diffuse spot of light only. But the big telescopes catch every bit of this light down to very faint sources. Now the astronomers do what they like best: they count. They count all these galaxies down to the tiniest speck of light and add up the radiation energy. Then they try to estimate the volume of space which contains the sources they have counted.

The positions of the galaxies on the sky, and their distances have to be known for that.

The distances are estimated from the Hubble relation, and the redshifts and positions can easily be measured. Thus the spatial volume emitting the radiation is known, and therefore the radiation energy per volume, called luminosity density, can be computed.

One step is still missing to find the mass density of the luminous objects: Radiation must be connected to mass.

The theory of stellar evolution tells us how much light a star of a certain mass emits, and from precise observations in the solar neighborhood we know how the stars are distributed according to mass. There are very many stars with a small mass, and only a few with a big mass, because the small ones live long, the big ones live a short time. This fact can be expressed quantitatively as the mean ratio of mass and luminosity for stars.

Multiplication of this ratio with the luminosity density results in a value for the mean mass density of the luminous matter. About half a percent of the critical density is the estimate to date. Expressed in terms of a density parameter Ω_* (* stands for star)

$$\Omega_* = 0.005.$$

There are, however, various possibilities for errors in this estimation: The galaxies chosen may not have been the most typical objects representing luminous matter, and also the Hubble constant itself is measured with some uncertainties. But the observers have counted galaxies in many different volumes – with somewhat different results – but nevertheless found that this value for Ω_* is quite reliable. It could be twice as big, but there is little doubt that the luminous matter reaches at most 1% of the critical density of the cosmic substance.

2.3.4 Dark Matter in Galaxies

In spiral galaxies the stars are arranged in a flat disk which rotates around the center. Astronomers have succeeded to measure rotational velocities at large distances from the center far outside of the luminous disk. They achieve this by observing the radio emission of clouds of neutral hydrogen. It turns out that mass is not concentrated near the central region, but that there is a nonluminous component of matter extending much further out than the visible light. Elliptical galaxies, which appear as luminous small disks without spiral arms do not rotate as a whole, but they also show evidence for dark matter, if the irregular velocities of their stars are analyzed. The mass in galaxies thus contributes somewhat more to the overall density than just the mass in stars. It reaches about 1.5% of the critical density,

$$\Omega_{Gal} = 0.015.$$

2.3.5 Dark Matter in Clusters of Galaxies

The galaxies are mostly bound in larger structures, especially dense assemblies of many hundred galaxies, so-called clusters. Their typical size is about ten million light-years (3 Mpc). These clusters are considered to be objects held together by their own gravitational force. The velocities of galaxies in clusters are, however, so high that the clusters would fly apart, if not additional dark masses existed which held them bound together.

Measuring the velocities of the galaxies, and applying Kepler's law to clusters, enables one to write down a mass balance. This forces one to accept a high fraction of dark matter in clusters. The nonluminous matter in a spherical halo around the galaxies cannot account for that. About ten times as much dark matter is necessary. This result is supported by further observations, such as the X-ray emission of galaxy clusters. A hot intracluster gas of a temperature of about 100 million degrees probably is responsible for the X-ray emission. A hot gas like that would simply evaporate from the cluster, if it were not bound by the gravity of additional, nonluminous masses. The quantitative estimates give a value for the density in agreement with the density derived from the motion of galaxies in clusters.

Many galaxy clusters act like a gravitational lens, that is they deflect light rays passing through the cluster which come from galaxies farther away from us to the cluster. The mass distribution in the cluster distorts the image of the source galaxy, and the analysis of the distortion allows us to reconstruct the mass distribution. These measurements indicate the same high fraction of dark matter in galaxy clusters.

All these data indicate that matter clumped on the scale of galaxy clusters adds up to a contribution to the total density of 15%,

$$\Omega_{cl} = 0.15.$$

The uncertainties still are considerable, and we should not exclude values higher by a factor 2. Thus a cautious estimate is

$$\Omega_{cl} = 0.3.$$

The astonishing result in any case is that dark matter is the dominant form of matter. There is about 30 times more dark matter than luminous matter. The normal matter, the chemical elements known to us, the "baryonic" matter as the physicists say, accounts for only 5% of the critical density, as we shall see below. There must be dark matter which is of a kind yet unknown. What could this unknown dark matter be?

2.3.6 Nonbaryonic Dark Matter

Astronomical measurements, and especially the analysis of the cosmic microwave background which will be discussed below, furnish many indications for the existence of dark matter which consists, apart from a small contribution of normal matter, largely of nonbaryonic matter. The elementary particles forming this dominating component of the matter are not yet known. We are familiar with neutrinos as representations of that species, but their mass is too small to contribute the required fraction of dark matter, although they originated in large number during the early epochs of the cosmos. In the Sun's interior neutrinos are produced continuously, and we meet them all the time without noticing it: They reach Earth in a steady flow and pass right through, also through us – about 100 billions of neutrinos per square centimeter, and per second! We do not feel them, because neutrinos interact only very weakly with matter. Even passing the big detectors in the underground mines of Kamioka (Japan) or Homestake (America) in tons of water they suffer only one collision per day on average.

Cosmologists take the neutrinos as examples, and postulate the existence of hypothetical particles which react weakly with normal matter like neutrinos, but which are much more massive. Up to now such particles have not been detected, although several experiments in underground laboratories have been set up to look for them. There are, on the other hand, a number of theoretical candidates. A favorite among them is the "neutralino," a particle without electric charge and with a mass of a few times the mass of the proton.

2.3.7 Galaxy Formation

The strategy in the theoretical modeling of galaxy formation has been to compute first of all the structures forming in the dark matter. This seems reasonable since there is evidence for about 10–100 times more dark than luminous matter in the cosmic structures. In a second step then the normal baryonic matter is distributed in the gravitational potential wells of the dark matter. The simulation of gas and dark matter together requires enormous computing power, and is only carried out in specially selected cases.

Such numerical and analytical investigations of cosmic structure formation are a major research topic of groups all over the world.

2.3.8 Dark Halos and Luminous Galaxies

Theoreticians have gained a lot of insights into the properties of structures formed by dark matter particles.

Although dark matter particles experience only their mutual gravitational force, the computation of their possible configurations is not quite easy, because the scientists want to follow the evolution of millions of particles to see what kind of structures are forming. This requires extensive computer simulations and numerical skills.

Some of the principal aspects can be clarified without too much mathematics. Consider a spatial volume in the expanding universe which contains a bit more mass than the average.

Under the influence of its own gravity this volume lags a bit behind the general cosmic expansion. Therefore matter is becoming less dense also in this region, but not to the same extent as outside. The contrast to the exterior region will increase in the course of the cosmic expansion and at a certain time become so large that within the volume considered the self-gravitation dominates. Then this clump of material separates, does not expand any longer, but collapses, and participates in the cosmic expansion as one whole object. This condensation of dark matter is called a "halo." The dark matter halo collects some normal matter which forms stars, galaxies, and galaxy clusters.

Let us assume for simplicity that the halo was spherical. Then after separation the density in the halo is about 180 times larger than the average cosmic density ($18\pi^2$ in a $K = 0$ model). Actually halos should rather be elliptical as numerical simulations have shown.

Figure 2.13 shows a section of a numerical simulation containing 16,777,216 particles of dark matter in a cube with an edge of 300 million light-years. The brightly colored areas are those with a very high density, and here you would expect the formation of luminous objects. There can be discerned various large-scale structures of high density like sheets or filaments, and also extended almost empty regions. All these qualitative features agree completely with astronomical data.

In Fig. 2.14 some results of the Las Campanas Redshift Survey are displayed, about 30,000 galaxies with redshifts up to 0.2. According to the Hubble law of expansion these galaxies have flight velocities of up to 60,000 km s^{-1}. Their proper velocities of a few 100 km s^{-1} are in comparison quite insignificant. Thus one may use Hubble's law to estimate the distances to the galaxies of the survey, and taking the positions on the sky into account, one arrives at a three-dimensional picture of their distribution. Figure 2.14 contains galaxies selected from three bands across the sky of 6° latitude extent each and covering about 120° in longitude (the so-called right ascension). In this wedge diagram the galaxies are plotted according to their longitude and their redshift, while the latitude coordinate is compressed. The observer is situated at the tip of the wedge.

The spatial distribution appears extremely inhomogeneous. Almost all galaxies are in extended thin sheets which surround like a skin large, empty volumes (voids). The picture

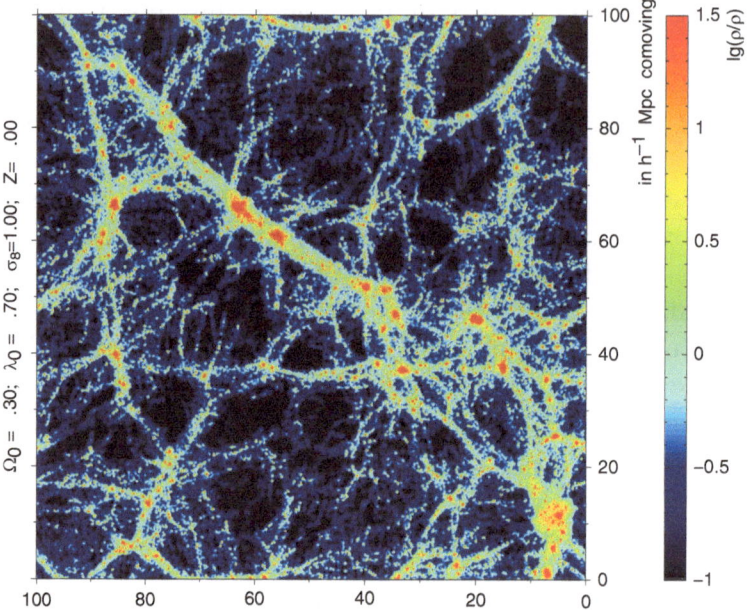

Fig. 2.13 A cross section through a cubic volume of a numerical simulation with dark matter particles shows similar condensations and voids as the observed galaxy distribution. *Bright (red)* areas mark a high concentration of particles, that is a big mass with a strong gravitational attraction. *Dark areas* do not contain particles, and therefore also no galaxies. The real volume represented by this simulation has a typical dimension of 300 million light-years. Not only by eye impression, but also in quantitative statistical measurements these simulations agree well with astronomical observations

of a spongelike pattern with galaxies situated in the thin walls of almost spherical voids seems adequate. Rich clusters of galaxies are located in places, where several walls come together. Quantitative comparisons must be done by employing a detailed model of galaxy formation. The crucial point is how to place galaxies in halos of dark matter. This is, of course, fully determined by the basic physical processes, but it is not yet possible to carry out the full-scale computations necessary to describe the complex behavior involved in the heating and cooling of the gas, the formation of stars, and the stellar explosions. Therefore the cosmologists test various recipes of how to populate halos with galaxies. Depending on the mass and history of a halo it may contain massive or very small, a few or many galaxies. The models are compared to the data in extensive

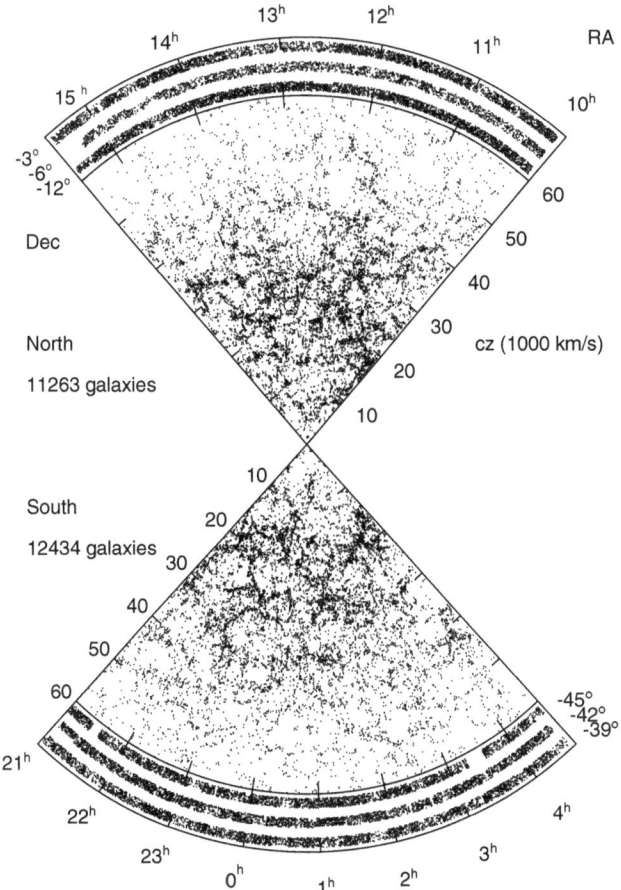

Fig. 2.14 Modern fast methods of measuring redshifts make it possible to undertake a cartography of the spatial distribution of the galaxies. To that end a galaxy catalog which lists positions on the sky of all the galaxies in a certain section of the sky and down to a limiting apparent brightness is used, and the redshifts of all the galaxies in it are measured. In this figure all the approximately 30,000 galaxies of the "Las Campanas Redshift Survey" are plotted in a wedge diagram of redshift against a sky coordinate. Only galaxies from three small bands across the sky are plotted, and the second positional coordinate is suppressed. One can clearly see characteristic features of the distribution: In a cell- or spongelike structure galaxies are localized in "walls" which surround large, almost empty volumes. The observer in this diagram sits at the tip of the wedge and surveys an angular section to the north as well as to the south

quantitative, statistical analysis. It turns out that the distribution of the galaxies in space, as well as their mean velocities, is well reproduced in the models, if the initial density fluctuations and the cosmological model are chosen adequately. Best fits are achieved for models which are at the critical density $\Omega = 1$, with 30% contributed by matter, and 70% by a cosmological constant. Similar values are obtained from an analysis of the anisotropies of the CMB (see below).

Not only in its present state, but also at earlier epochs can the galaxy formation model be tested, because meanwhile even at large redshifts many galaxies have been detected. All these tests show that the theoretical scenarios provide a reliable description of structure formation, even if not all details are correctly implemented as yet.

The earliest condensed hydrogen clouds are observed with redshift between 6 and 10. Such an early epoch can be reached only with the biggest telescopes available, and even then only a few spectral lines can be registered, no images. But in these spectra one finds not only the lines of hydrogen and helium, but also signatures of heavier elements. Even in these early epochs there must have existed stars which had after their explosion enriched the cosmic material with traces of carbon, oxygen, and magnesium. At redshifts around 3 astronomers find fully evolved galaxies shining in starlight in large numbers. Each galaxy is thought to lie inside a halo of dark matter.

For many years now the halo of the Milky Way has been investigated in large surveys. Astronomers are looking for a special phenomenon, the "microlensing" effect: The bending of light due to gravity can lead to a significant increase in the brightness of a distant star exactly in the case, when the straight line from the observer to the star just grazes the edge of a massive dark body in the halo. If the halo consisted of such objects which have received the pretty name "MACHOs" (massive compact halo objects), then some distant stars would occasionally brighten for a short time. The halo objects are not visible, but the effect of their gravitational potential on the light rays coming from a star outside. The light rays are deflected and bundled such that the passage of the MACHO leads to a brightening and subsequent completely symmetrical darkening of the star.

Millions of stars in the Large Magellanic Cloud have been surveyed now for about a decade. Several tens of microlensing effects have been observed. The conclusion is that about 30% of the halo mass lies in small, nonluminous celestial bodies. The remaining 70% of the dark matter of the halo of the Milky Way are supposed to be nonclumped exotic elementary particles.

2.3.9 Stars and Elements

The first stars formed in the condensing clouds of hydrogen and helium which we find as the predecessors of galaxies in the universe at redshift 6 and larger, i.e., when the universe had about one-tenth of its present size, and a density about thousand times bigger than now. In the interiors of these first massive stars the chemical elements heavier than helium – carbon, oxygen, and iron – were brewed. Every carbon or oxygen atom in our body has gone through several generations of stars, expelled into interstellar space in supernova explosions, recycled in the evolution of a new generation of stars, until it finally ended up on the Earth, when the solar system was formed. We consist literally of "stardust." The generations of normal stars which formed in a medium, where the heavy elements had been available already, with planetary systems around them, are a consequence of evolutionary processes which began in the early universe.

Why does this take billions of years? Well, the force of gravity is very weak, and thus it needs a long time to condense massive objects out of the cosmic matter which is blown apart by the tremendous momentum of the original cosmic explosion. The steady flow of energy from a star like our Sun, and the solid surface of a planet like Earth with its concentration of heavy elements finally provide favorable conditions for the origin of complex biological structures.

2.4 The Cosmic Microwave Background (CMB)

The big-bang model provides a simple and obvious explanation for the CMB as the relic radiation from a hot early phase of the cosmos. Therefore the CMB is considered as one of the important supporting pillars of this cosmological model. Apart from the more indirect arguments connected to the synthesis and present-day abundance of the light elements, there is no further experimental evidence of the early cosmic history.

Alternative cosmological models are sometimes being brought into the discussion, but they all fail to reproduce the uniformity and the ideal black-body spectrum of the CMB. Because of its enormous impact on our knowledge of the universe, the properties of the CMB should be discussed in some detail. I want to do this in the following.

The CMB is important, because its smoothness supports the idea of the uniform and homogeneous cosmological models, and also, because the small anisotropies of the CMB allow us to determine precisely the parameters of the models, such as the energy and matter density. Thus the CMB presents us with an independent approach to cosmic data besides the astronomical observations of stars and galaxies.

Within the framework of the FL models, we can trace the history of the cosmos to the past. As we reach earlier and earlier times, we find that the cosmic radiation field contained sufficient numbers of energetic photons to ionize all hydrogen atoms, i.e., to prevent the hydrogen nuclei, the protons, from forming an atom by binding an electron. This was still the state of affairs, when the average CMB temperature was about 3,000 K. At that time, about 400,000 years after the big bang, about one out of every billion photons in the CMB had an energy greater than the energy of ionization of a hydrogen atom, of 13.6 eV. That was just what was needed to keep the hydrogen nuclei separated from the electrons. Matter was composed of a rather uniform hot plasma. Stars and galaxies did not yet exist in that early epoch.

But due to the expansion the system cooled, and gradually first forms appeared in the primeval soup. At temperatures below 3,000 K the free electrons started to combine with the atomic nuclei to form hydrogen and helium. During this stage of "recombination" – as it is called inappropriately, because in fact hydrogen and helium atoms formed for the first time in cosmic history – the universe became transparent, the scattering of photons on electrons was strongly reduced. This happened within a short time span, but not suddenly – the process of "recombination" took about 40,000 years. The spectrum of the CMB does not show any features from this phase. No deviation from a Planckian spectrum (Fig. 2.10) with a temperature of

$$T_\gamma = 2.728 \pm 0.002 Kelvin$$

was found, and this is another, very beautiful fact in favor of the simple cosmological big-bang models: Even during the 40,000 years of recombination the temperature of the radiation and the photon energy must have followed perfectly the equations describing the FL models. Thus the shape of the Planckian spectrum has remained unchanged, while the intensity of the CMB (its energy density) decreased in proportion to the fourth power of the temperature.

2.4.1 Acoustic Oscillations in the Early Universe

Much more can be read out from the CMB. Mass concentrations of the dark matter had been forming already before the recombination epoch albeit with a very weak density contrast. The tightly coupled plasma of photons and baryons (essentially hydrogen and helium nuclei) followed these condensations, but the desire of the baryons to clump together was counteracted by the photon pressure which drove these plasma clouds apart. The competition of these two forces caused the plasma condensations to oscillate – a behavior analogous to sound waves. The largest oscillating plasma cloud had been crossed just once by a sound wave during the time interval from the big bang to the recombination time. Bigger clouds did not have enough time to develop a pressure counteracting gravity and just followed gravity by contracting slowly. Smaller clouds oscillated with higher frequency. All the oscillations were perfectly synchronized by the big bang. Contraction of the plasma condensations increased the density and heated up the photon gas, expansion decreased the density and cooled the photon gas. At the epoch of recombination the photons left the plasma clouds. Now they appear with slightly different temperatures in the detectors of the astronomers. The temperature fluctuations show up as hot and cold spots in the CMB sky maps.

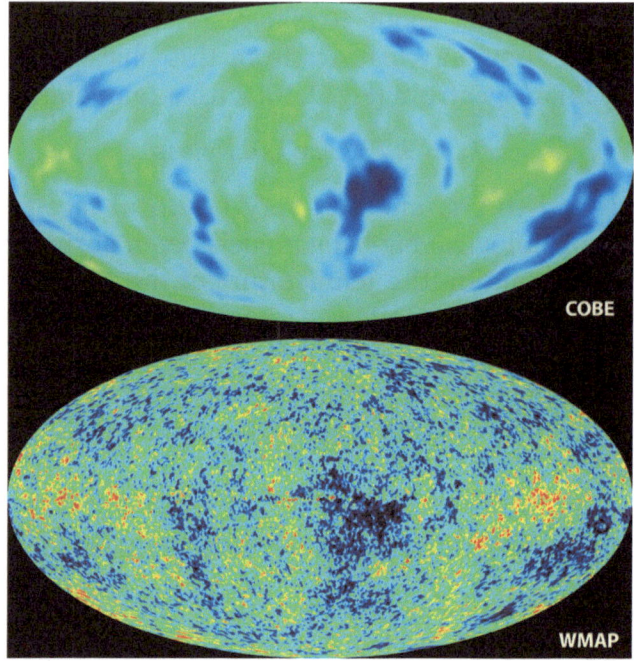

Fig. 2.15 A comparison of the sky maps obtained from the measurements of temperature fluctuations of the CMB by the satellites COBE and WMAP clearly demonstrates the improved resolution of the WMAP instruments. One also sees that specific spots of higher temperature in the COBE image (colored *yellow*) have corresponding spots in the WMAP map (*courtesy of the WMAP collaboration*)

In 1992 the first successful measurements of structure in the CMB have been carried out with NASA's satellite COBE. The sky maps obtained showed hot and cold spots on the sky with relative amplitudes of $\frac{\Delta T}{T} \simeq 10^{-5}$ (cf. Fig. 2.15).

The instruments aboard COBE had rather low angular resolution, the satellite was too "short-sighted" to recognize small structures, the angular extent had to be about 7° before a spot on the sky would be identified as a measuring point. If COBE had been looking down onto the Earth, then the whole of Bavaria would just have been one measuring point (cf. Fig. 2.16). The variations in intensity which would mirror the seeds of galaxies and galaxy clusters are expected to be on scales well below 1°.

In 2001 the satellite MAP was launched by NASA. MAP surveys the CMB sky with an angular resolution of about 15 arcmin

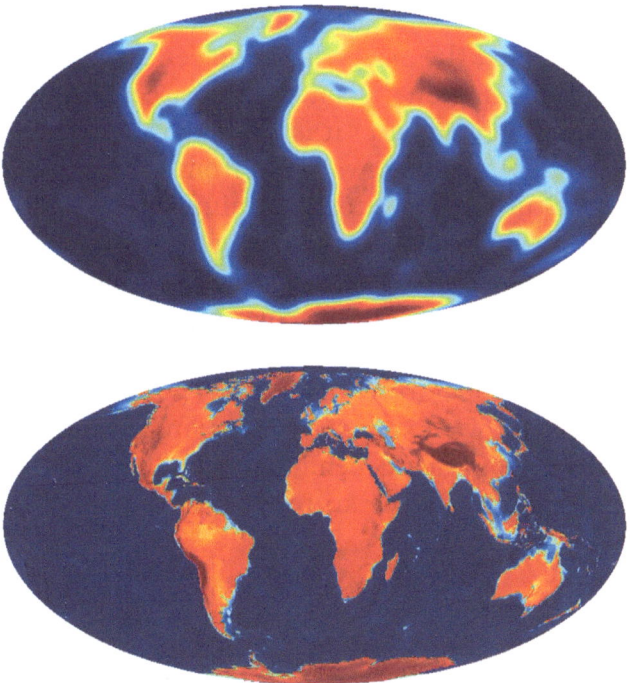

Fig. 2.16 The better resolution of WMAP observations can be illustrated by a fictitious view of the Earth by COBE (*left*) and WMAP (*right*). Bavaria would be one pixel for COBE, while for WMAP Munich would be one measurement pixel (*courtesy of M. Bartelmann, University of Heidelberg*)

in a range of wavelengths from 3 mm to 1.5 cm. The satellite was later renamed WMAP to honor David T. Wilkinson, a pioneer of CMB research, who passed away in September 2002.

The European satellite PLANCK has been launched in 2009. It has an angular resolution of 5 arcmin and covers a significantly wider range of wavelengths from 0.3 mm to 1 cm. The angular resolution of PLANCK is good enough to retrieve a major fraction of the spectrum of acoustic oscillations. Temperature fluctuations of the order of a microkelvin can be registered.

Both satellites measure besides the intensity of the CMB also its polarization properties which opens an additional window on cosmological parameters. Such measurements have been made possible by the development of new radiation detectors which are cooled down to temperatures of about 100 mK.

WMAP and PLANCK are located at a point outside of the Earth's orbit around the Sun, where the centrifugal and the gravitational forces acting on the satellites just cancel each other. At this "outer Lagrangian point" it is possible to orient the satellites such that they always look away both from the Earth and the Sun. In that way disturbing radiation is minimized.

Meanwhile the observational data gained with WMAP for the first 5 years of observation have been analyzed. The sky map of the CMB agrees well with previous experiments (cf. Fig. 2.15). As a result of these measurements astronomers can construct a power spectrum of the temperature fluctuations (Fig. 2.17).

The graph shows a sequence of maxima and minima of the temperature fluctuations depending on the angular scale in the sky over which the temperature has been averaged. The first

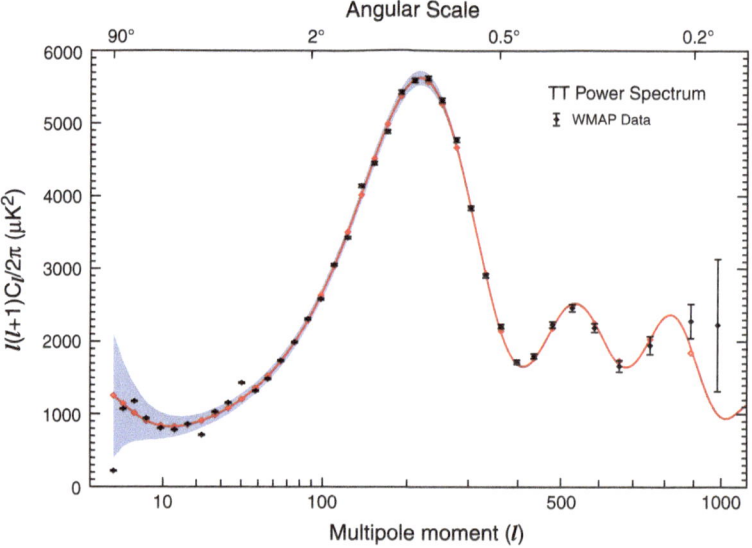

Fig. 2.17 The graph in this figure is the power spectrum of CMB anisotropies. It shows the square of the temperature fluctuations expanded in terms of multipoles. In a more intuitive way one might describe this as the square of the temperature difference between two small pixels on the sky separated by a certain angle, and then averaged over all pixel pairs. Many cosmological parameters can be read off from the shape of the curve, and its dependence on multipole index (ℓ) or angle $(\sim 200/\ell$ degrees). The regular sequence of maxima is as expected from the theoretical models of structure formation. The location of the first maximum at $\ell = 200$, and an angle $\simeq 1°$ shows that the spatial curvature of the cosmos is zero (courtesy of WMAP collaboration)

maximum corresponds to the largest acoustic oscillation – its wavelength is the distance covered by a sound-wave in the time span between the big bang and the era of recombination. This distance appears on the CMB sky as a prominent signal with an angular extent of about 1°. This result tells us interesting facts about the structure of space: The viewing angle of a given length is determined by the curvature of space. The same length viewed in a space with positive curvature appears at a larger angle than in a zero curvature space, and at a smaller angle, when the curvature is negative:

The measured value of 1° means that the spatial curvature is zero, i.e., the Universe obeys Euclidean geometry – it is as simple as possible, geometrically. Curvature zero also means that the total mass and energy density Ω_{tot} reaches the critical value. Exact analysis results in

$$\Omega_{tot} = 1.00 \pm 0.03.$$

Only a small positive or negative curvature (a 3% deviation of the density parameter) is still acceptable within the limits of accuracy of the measurements.

The acoustic oscillations are a sequence of expansions and contractions, and a higher fraction of baryons causes a deeper contraction. The ratio of the amplitudes permits to derive (for a Hubble constant of 70)

$$\Omega_B = 0.044 \pm 0.003$$

for baryonic matter, and

$$\Omega_{CDM} = 0.21 \pm 0.03$$

for dark non-baryonic matter. These values are in excellent agreement with other astronomical measurements.

Baryonic and dark matter together reach only 26% of the critical density $\Omega_{tot} = 1$. Therefore there must be a further component of the cosmic energy density which balances this deficit. This component must be distributed uniformly; it must not show clumping on scales of galaxy clusters or below. It seems necessary to postulate a uniform cosmic energy density Ω_Λ with a range of values around 74%:

$$\Omega_\Lambda = 0.74 \pm 0.03.$$

A best fit to the CMB data yields values for the cosmic energy density components of

$$\Omega_{tot} = 1, \Omega_{\Lambda} = 0.74, \Omega_{CDM} = 0.21, \Omega_B = 0.05,$$

(see also Fig. 2.18).

2.4.2 Dark Matter and Dark Energy

Although the physicists have no direct experimental evidence yet of the nature of dark matter, there are many indications from astronomical observations that it resides in galaxies and in clusters of galaxies. Supposedly, it consists of elementary particles which have not yet been detected, but which are sought after in several experiments.

Even with dark matter there remains a gap of about 74% in the cosmic energy balance. Physicists are inclined to balance the deficit by the energy of a suitable field or by the energy of the vacuum, the ground state of the world. This reminds us of

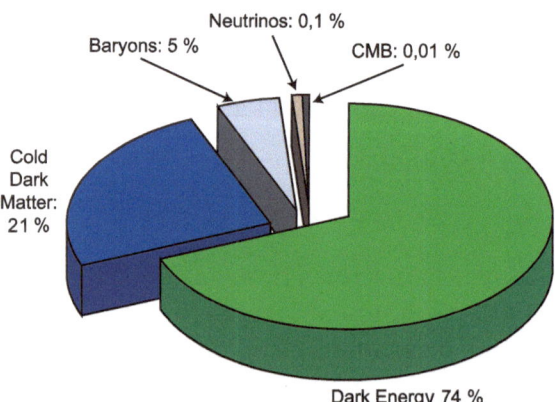

Fig. 2.18 The remarkable composition of the cosmic substrate is displayed in this diagram. Only about 5% of the cosmic matter and energy density are known. The sector inscribed "baryons" designates the fraction of matter known to us, the elements of the Periodic system. The small amounts contributed by the CMB (marked "CMB") and by neutrinos are also shown. The big majority is unknown: dark matter (21%) and dark energy (74%)

the futile attempt of Einstein to construct a static universe by the introduction of a cosmological constant. Similar to such a quantity, an almost constant field energy would accelerate the cosmic expansion, in contrast to the massive bodies in the cosmos which would decelerate the expansion due to their mutual gravitational attraction. The missing 74% of the cosmic energy density have been named "Dark Energy" (commonly written in capital letters, a custom I will not follow, because I consider the name "dark energy" a bit misleading). The dark energy would grow proportional to the spatial volume during expansion – its density would remain constant. A gas of particles on the other hand has an energy which stayed constant in an expanding volume; its energy density would shrink inversely proportional to the volume. Such a different behavior also has the consequence that the dark energy, small as it has been initially, will dominate in the course of time.

What then is dark energy? Quantum theory might help us in understanding this quantity as the energy of the vacuum. From the point of view of quantum theory empty space is a complex structure of interwoven fluctuating fields which cannot be observed, but which contribute to the energy of the ground state nevertheless. Some of these contributions can be estimated by theorists quite well, but they compute values which exceed the observational number by 60–120 powers of ten. Other contributions which cannot be computed (so far) might perhaps balance this value, but the balance must be incredibly accurate: down to the 120 first digits after the comma. It is one of the great mysteries of physics how this might be achieved.

It is absolutely remarkable that here a fundamental problem of quantum theory has become apparent through astronomical measurements. In all approaches to theories of elementary particles vacuum energies arise, but obviously they do not have a gravitational effect.

A side remark by the brilliant theoretical physicist and Nobel prize winner Wolfgang Pauli illustrates the problem nicely: A few years after the proposal of the theory of general relativity by Albert Einstein, Pauli calculated the radius of the universe under the assumption that the zero point energy of the electromagnetic field determines the value of the cosmological constant. He found that the radius of this universe would be smaller than the distance from the Earth to the Moon, in other words light rays in this cosmos would be deflected so strongly that we could not even see the Moon. This demonstrates the big discrepancy between theoretical predictions and the real situation.

The final theory, if it will ever be found, must also explain why the energy density of the vacuum is gravitationally inactive, contrary to all other kinds of energy densities. There is, of course, hope that a theory of everything, especially a unification of quantum theory and the theory of gravitation, will improve our understanding of these questions decisively. At the moment we just have to acknowledge the problem. We may also take note of the fact that for experiments in the laboratory only differences of energy count, and therefore this difficulty does not arise. Only when we consider the universe as a whole, the absolute value of the energy density plays an important role.

Thus we can only attempt to give a more precise mathematical description of our ignorance, perhaps by describing the dark energy as the energy of a field with the right properties. The beautiful name "quintessence" has been coined for such a designer-made field. But it remains actually a mystery why dark energy exists at all, and why it determines just now the cosmic expansion. If the dark energy remains constant, the cosmic expansion will continue forever and forever accelerate. But to link dark energy to the idea of field energy offers the interesting possibility that in the future the field will change with time and surprising new turns in the cosmic evolution may occur.

2.4.3 An Effect of Five Percent

Several remarkable insights follow from the study of the expanding cosmos. Evidently the big-bang model is a convenient framework to accommodate the cosmologically relevant observations in a model of cosmic evolution. The synthesis of the elements, the formation of structures in the universe can be explained without any great effort by such a model. To be sure we have to swallow a bitter pill with all that – or empty a whole glass of vermouth – because 95% of the cosmic substrate are unknown to us. We ourselves, the things around us, the planets and stars, are only a marginal phenomenon, a five percent effect, in the universe. Why is that so? Can we try to find some deeper explanation for it?

It seems to me that to this end we have to leave the area of secure knowledge, and to look at some speculative ideas about the earliest epochs in the universe.

2.5 The First Second

About one second after the big bang we can describe with some confidence the physical processes in the early universe, because then the conditions are not too different from those known from terrestrial laboratory experiments, and the known laws of physics should hold. But the first fractions of a second after the big bang are the area of more or less well-founded speculation. Close to the big bang in the standard model thermal energies are far above the energies reached in terrestrial particle accelerators. Finally, in the initial fireball temperature and density grow beyond any limit. Right at the big bang temperature, density, and curvature become infinite. The cosmological model loses its ability to describe the situation in terms of acceptable physics. Even Einstein's theory of gravitation fails at the singular big bang. It is admirable nevertheless that the theory exhibits its limits of validity on its own.

The popular question "What was there before the big bang?" leads beyond this singular boundary, and is by physicists often felt to be "not allowed," since time originated with the big bang itself, and therefore an earlier moment of time cannot exist, at least not in this model. But it seems legitimate to ask, whether for the big-bang model preliminary conditions of some kind can be imagined.

Very likely, the description of the cosmos as a classical space–time must be given up, if one wants to find out more about the beginning. Very close to the big-bang singularity the whole universe becomes in a (somewhat fuzzy) sense a quantum object. Without a unified theory which encompasses gravitation and quantum mechanics, all attempts at a more detailed description of the beginning must therefore be counted as speculative exercises. As long as such a theory is not available, one may try a more modest approach, and investigate the consequences of connecting a quantum description of matter and radiation with the classical space–time of the cosmological standard model.

It is fascinating to play around with the possibilities of cosmology and particle physics, and ask what kind of minimal structure had to be imprinted on the big bang itself, and which properties might have evolved out of physical processes.

The conceptions of elementary particle physics which come into play here will be discussed in detail in Chap. 3. Here only some basic characteristic features will be mentioned. Nevertheless some important connections between cosmology and particle physics will be pointed out. A typical example is the problem of how to explain the ratio of the number of matter to radiation particles: A ratio of about 10 billion quanta of radiation per one particle of matter characterizes the present state of the cosmos.

This ratio means that in the early phases of the cosmos the hot primeval plasma consisted primarily of particles and antiparticles (same mass, but opposite sign of charge as the corresponding particle) in almost equal numbers, but with a tiny surplus of 10 billion plus one particles versus 10 billion antiparticles. In the course of the cosmic expansion the primeval plasma cooled, particles and antiparticles annihilated into radiation, and the small surplus of particles of one in a billion remained.

We owe our existence to that tiny effect! Now one investigates the question whether this small asymmetry can be produced by the interactions of elementary particles from a completely symmetric initial state. Some more recent theoretical considerations make it quite plausible that this could happen during the phase transition, when the electroweak force splits up into the weak and the electromagnetic force, about 10^{-10} s after the big bang.

2.5.1 The Inflationary Model

The inflationary universe model has been the most popular scenario during the past 25 years, whenever cosmologists tried to describe the situation as close as possible to the singular big bang. What would happen, if right at the beginning not radiation and matter, but the energy of a field determined the dynamics of the cosmos? Physicists in Japan, the Soviet Union, and the USA asked themselves this question independently in 1981. They all found that in this case a dramatic change of the cosmic expansion would take place, an extreme acceleration of the expansion, where the separation of two particles would double every 10^{-35} s. During the tiny time interval between 10^{-35} and 10^{-33} s after the big bang – characterizing such models – the distance between particles would have been increased by the factor 10^{29}, while in a radiation-dominated FL model only a growth by a factor 100 would have occurred. The driving power behind such an "inflation" might be the energy of a scalar field. The existence of such a field was proposed originally in analogy to the designs of a unified theory of elementary particles, known under the acronym GUT ("Grand Unified Theory," cf. Chap. 3). In GUTs there are fields of this kind, so-called Higgs fields, which are introduced to describe the symmetry breaking responsible for the transition from a single fundamental force to the hierarchy of weak, electromagnetic, and strong forces observed today. Such designs suggest that the early universe was full of scalar fields, although up to now a scalar field has not been found in any experiment. The universe might have evolved from an initial phase of high symmetry with a high energy density of the scalar field in the course of continuing expansion and cooling to a state of low field energy density, and lower symmetry. If the phase transition from the symmetric to the asymmetric state occurs not immediately, but gradually and delayed, then the energy difference between the states of the scalar field may influence the expansion. It may even dominate over other thermal energies. The high-energy, highly symmetric initial state has been named "false vacuum," to indicate that it is not permanent, since the field will at last settle into the favored configuration of lower energy.

In a schematic and intuitive way we may illustrate this transition, the "symmetry breaking" as in Fig. 2.19.

Initially a small sphere lies on top of an ideal Mexican hat in the gravitational field of the Earth. The gravitational force is directed parallel to the axis of symmetry of the hat (a good approximation to the real situation on the Earth's surface). Therefore a rotation around this axis does not change anything, the system is rotationally symmetric. If the sphere rolls down into the brim of the hat and lies there at some specific location, the rotational

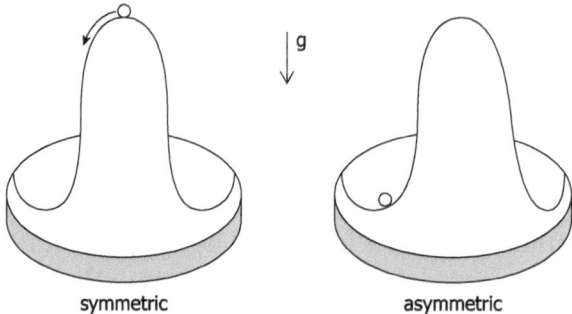

symmetric   asymmetric

Fig. 2.19 A mechanical example for a symmetric state is a small spherical ball on top of a Mexican hat in the gravitational field on the Earth. The force of gravity points along the axis of the hat, and therefore this configuration is rotationally symmetric. Rotations around the axis of the hat do not change anything. But if the ball rolls down, and comes to rest somewhere inside the brim of the hat, the rotational symmetry is gone. In the inflationary universe model the symmetric state is interpreted as a "false vacuum" with a high energy density. The asymmetric state is the "right vacuum," where the scalar field has its lowest energy

symmetry is lost. Between the brim and the top of the hat there is a gravitational potential difference, the analog to the energy of the scalar field in the inflationary model.

But the analogy cannot be carried any further, because the energy density of the false vacuum has a remarkable property which is quite different from the behavior of normal matter. While the energy density decreases in an expanding volume filled with matter, the false vacuum keeps a constant energy density, does not thin out in an expanding volume. In fact, it is the ground state of the world, the "vacuum," which is determined by the value of the Higgs fields.

Even during the cosmic expansion the desire remains to stay in this ground state. This property is caused by the strange relation between pressure and density which holds for the vacuum, for which the pressure is equal to the negative energy density. Enlargement of a volume is then equivalent to work against a negative pressure, i.e., to gain energy. This special property of the false vacuum leads to the effect that in epochs when the energy density of the false vacuum dominates, the cosmic expansion is accelerated. This energy density acts like a repulsive force on the masses!

During the short time span of inflation all the objects that were there before are thinned out dramatically, their density becomes negligible. The temperature also goes down by the inflationary factor. The curvature of space–time is smoothed out like the wrinkles in a balloon, if it is blown up.

The inflationary phase persisted, as long as the scalar field stayed in the state of the false vacuum. It ended, when the field had reached its minimal energy. In the final phase the energy density of the false vacuum was transformed into a gas of hot radiation and particles. From this time onward the universe evolved as described by the standard model of cosmology, but with initial conditions which had been determined at least partly by physical processes. This "second beginning" requires that all matter, energy, and entropy of the observable part of the universe were created by the decay of the false vacuum.

You have every right to ask whether these doubtlessly extraordinary aspects of the first few fractions of a second have any effect on the present state of the world. Yes, they do, quite astonishingly in several respects: First of all, there is the problem of fine-tuning of the standard big-bang model which cannot be easily reduced to more fundamental, simpler properties.

Thus, the mean density must be close to the critical value, $\Omega = 1$, for the universe to exist for a sufficiently long time, and to build up enough structure. At very early times, close to the big bang, the density must be very precisely close to the critical value. Small differences would lead to early collapse, if the density was larger than critical. On the other hand, a density smaller than critical, would lead to rapid expansion, a rapid thinning out of matter, such that structures like galaxies or stars could not form.

Another difficulty of FL models is the existence of causally disconnected regions: Different space–time points can be connected by light signals only, if their separation is small compared to the size of the universe. We may try to understand this in the illustrative analogy of the balloon surface: The region connected causally to a given point, the horizon of that point, can be represented by a circle on the balloon surface. The radius of this circle grows with the square of the radius of the balloon, when the case of the radiation-dominated universe is considered. Looking to the past in this picture, we see the balloon shrink, and the horizon shrinking even faster. Any length scale on the surface of the balloon changes just proportional to the radius, and therefore any particular point which is now inside the horizon has been outside at sufficiently early times. Therefore the causal structure at early times is weird: Less and less of space is contained within the horizon of each point, the space–time splits up into a growing number of causally disconnected regions, until right at the big bang each point is completely on its own. This "horizon problem" has an especially awkward significance, when we apply it to the CMB. Observing the sky in opposite directions, we see the same CMB temperature. But at the time of recombination such regions have been separated by about 70 horizon lengths. How then, could the temperature be the same, so precisely?

The inflationary model solves all these problems by the huge expansion of space–time. The universe undergoing inflation has a curvature approaching zero, and a density close to the critical value. The horizon problem is solved, because the whole observable universe could have grown out of a tiny initial seed, a space–time bubble which was just a small part of one horizon. How big must this initial space have been? We can compute back from the present state with a temperature of 2.7 K, and a typical extent of 10^{28} cm to the epoch just at the end of inflation.

At that time the observable cosmos had a size of about 10 cm. Since inflation stretches all length scales by at least a factor 10^{29}, a dimension of about 10^{-28} cm for the initial seed of our universe would be sufficient. This is about one thousand times smaller than the causally connected volume at the beginning of inflation, at a time of 10^{-35} s. The causal length at that time is $ct = 10^{-25}$ cm.

The Russian physicist Andrej Linde, who now lives in Stanford (CA), has adorned this picture with a lot of imagination, and sketched a grand view of the universe consisting of disconnected, continuously emerging and decaying cosmic bubbles. According to Linde we are in one of these bubbles, a special one, because it provides acceptable living conditions.

This universe of bubbles is continuously changing, some parts experience inflation, others remain in the false vacuum with fluctuating scalar fields, but in total it is an eternal state without beginning and end. There is no problem with the origin of the universe, since it is not even clear whether an overall conception of time can be found for this bubbling chaos. The hypothetical inflationary model of Linde has the amusing property that the initial mass of the universe is tiny, of the order of the Planck mass, i.e., 10^{-5} g, like the mass of a small bacterium. Thus to create a complete universe like ours, only a small investment of mass or energy is necessary, at least according to this speculation. So much for Andrej Linde's scenario of "chaotic inflation."

Another very important success of the inflationary model is its prediction of small fluctuations of the energy density, a necessary ingredient for cosmic structure formation. The quantum fluctuations of the scalar field which are always present are stretched by inflation such that they attain astronomically relevant dimensions. Within detailed models a spectrum of inhomogeneities can be derived with the property that the mass excess in a given volume is decreasing proportionally to the length dimension of that volume. Data from the satellites COBE and WMAP confirm this prediction for the spectrum of CMB anisotropies.

Besides these points in favor of the inflation model, we must also mention some of its drawbacks. Especially the attempts to transform the scenario into a more precise mathematical model have met with difficulties again and again. I do not want to consider here in detail these more technical questions, but at least point out one fundamental problem: The inflationary expansion is driven by the energy density of fields acting in the early universe. We know, however, from our experience that the vacuum energy densities of the actual strong, weak, or electromagnetic interactions must not be gravitationally active, because typical energy densities are so large that contradictions to the astronomical observations would be obvious. Only a modest contribution of the order of the critical density can be tolerated – such as the dark energy derived from CMB observations. The energies of the inflation fields are larger by about 120 orders of magnitude.

Why should vacuum energies have dominated the evolution in an early cosmic phase, if now they cannot be allowed to act gravitationally at all? A good idea would be very desirable.

2.5.2 The Beginning

If you are not satisfied with the explanation given by the inflationary model, you have to investigate the initial conditions for the universe. My feeling is that this is not a question within the scope of physics, but rather a metaphysical one. Restricting physics to the explanation of phenomena within the universe saves us from a lot of difficult problems. We may, however, ask whether we should stop at the simple classical picture of the big bang, or whether we could not use arguments from physics to approach the origin a bit further.

A nonphysical answer has been given by St. Augustine in his "Confessions" (vol. 11): "To the question 'What did God do, before he created the world?' some might be tempted to answer: 'Then he created Hell for people, who ask such questions'."

A singular event like the origin of the world evidently makes the distinction between initial conditions and laws of physics obsolete. Even though, we would like to know in more detail why and how the big bang happened. Is there perhaps a quantum state, a kind of primeval vacuum, out of which the universe rises, like a bubble from the "primeval foam"? This definitely sounds metaphysical, at least in our present state of knowledge, where a theory unifying quantum physics and gravity is still missing.

A name for such a theory has already been proposed, however: "quantum gravity."

Even while quantum gravity is not yet here – or exactly then – one may indulge in speculations as to how a quantum state of the universe might be described. The English physicist Stephen Hawking has followed such inquiries intensely. He proposes to consider as possible models for the quantum cosmos only simply structured, smooth space–times; thinking in terms of our balloon analogy only a smooth balloon without wrinkles. Time does not exist in such a quantum universe. There is only a sequence of simple four-dimensional spaces – the four-dimensional surfaces of five-dimensional spheres. For illustration we can look at our balloon picture, where the surface is two-dimensional. Now try to add two more dimensions in your imagination! That is not easy, but worth trying. From this quantum cosmos our universe suddenly jumps out, and enters its temporal evolution with a finite volume from the start.

Those considerations are of principal interest, even though a nut-sized universe seems no less fantastic than a singular big bang. The question "what was before the 'primeval nut'?" cannot be asked because normal space and time categories do not exist in the quantum cosmos. Quite similarly, it makes no sense to ask for the longitude and latitude of a point outside of the Earth.

Following a different line of arguments the British mathematician Roger Penrose also argues that at the beginning the universe must have been a space–time of extraordinary smoothness and uniformity.

His starting point is an experience, we all have made every now and then: Most everyday occurrences are not reversible. A glass of water falling down from the table to the floor, splintering and spilling water, shows the normal and expected run of events. The reverse behavior, when a broken glass on its own became whole again, and jumped up onto the table, as in a backward running movie, would certainly leave us perplexed. The laws of mechanics allow this reversal in time. But actually things always happen by themselves such that an ordered state changes to a less ordered one. The notion of "entropy" is very helpful to understand this property of nature. Entropy is defined as a quantity which measures the amount of disorder in a system. An ordered system, like a crystal, has a low entropy, a gas of molecules bouncing around irregularly has a high entropy.

The everyday experience of growing disorder corresponds to the law of increasing entropy (the "second law of thermodynamics"). The numerical values for the entropy of a system result from the possible different positions and velocities for each particle subject to the fixed total energy and the volume occupied by the particles.

Penrose attempts to characterize quantitatively the entropy of the universe, rather of its observable part. The numerical estimates reach gigantic values, if besides radiation and matter the possibilities to produce entropy hidden in the gravitational field, in the wrinkles and curvatures of space–time, especially in black holes, are included.

The initial conditions for the universe, as we know it, represent just one out of $10^{10^{120}}$ possible configurations of the cosmos. Can we postulate a selection principle of such precision within the scope of physics? A strict smoothness condition as suggested by Penrose might be a possible approach. But can this be derived from the basic equations? This is still hidden in the darkness of the unknown and unexplained.

Physicists will be engaged for some time to come in explaining the big bang. The fun involved in speculations and the enthusiasm for conceivable scenarios makes cosmologists prone to believe that what is conceivable is already real. To quote Albert Einstein: "To the inventor the products of his imagination appear so necessary and natural that he sees them and wants them to be seen not as structures of his thinking, but as given reality."

All considerations about the first moments of the universe, about its initial state and conditions, belong to the empire of metaphysical speculation.

2.6 The Anthropic Principle

In a situation, where the explanations of physics for the origin of the world reach their limit, a chain of arguments has found widespread interest which is called "anthropic principle." The fact that intelligent life exists on the Earth means that the conditions for the origin of intelligent life must be fulfilled in the universe. This rather trivial, logical statement of a necessary consistency has led to remarkable, nontrivial insights.

Life as we know it, could not have originated, if the constants of nature were slightly different from their actual values. The strength of the attractive nuclear force is just enough to overcome the electrical repulsion between the positively charged protons in the nuclei of common atoms like oxygen or carbon. But the nuclear force is not quite strong enough to bind two protons together. The diproton does not exist. But, if the attractive nuclear forces were a bit stronger, the diproton could have been formed, and then almost all the hydrogen in the cosmos would have ended up as diprotons or higher elements. Hydrogen in that case would be a rare element, and stars like the Sun generating energy over a long period of time by the slow fusion of hydrogen into helium would not exist. On the other hand, with a weaker nuclear force it would be impossible to have larger atomic nuclei. If a star like the

Sun generating energy at a constant rate over billions of years is necessary for the evolution of life, then the strength of the nuclear forces must be within narrow bounds.

A similar, but independent numerical fine-tuning can be found with the weak interaction which in reality steers the fusion of hydrogen in the Sun. The weak interaction is about a million times weaker than the strong interaction responsible for the nuclear force. It is just so weak as to ensure a slow and uniform burning of hydrogen in the Sun. Stellar lifetimes would change dramatically, if the weak interaction was somewhat stronger or weaker, and this would make it difficult for life depending on sun-like stars.

Another numerical agreement concerns the mean distance between the stars which in our galactic environment amounts to a few light-years. Maintaining the view that the stars can have a decisive influence on human life is not necessarily an argument from astrology. We would not have any great chance of survival, if the mean distance between stars were ten times smaller, for example. In that case another star would have come close to the Sun with high probability during the past 4 billion years. If it came close enough to disturb the planetary orbits, the effect might be disastrous. It would be sufficient to push the Earth into a slightly more eccentric, elliptical orbit to make life impossible.

One could enumerate many more happy constellations of this kind: A sensitive balance between electromagnetic and quantum mechanical forces causes the variety of organic chemistry. Because of these fine-tunings water is liquid, chains of carbon atoms form complex molecules, hydrogen atoms build links between molecules. But a small change of the constants of nature can destroy all that.

These numerical coincidences are statements of the "weak anthropic principle" which generally expresses the opinion that our existence is only possible under specific conditions. The scientists, who proposed this principle, want to draw attention to the remarkable harmony between the structure of the universe and the necessary requirements of life and intelligence. Our universe satisfies these conditions, but the reason why it does so cannot be explained within the scope of present-day physics.

There are also those, who advocate a "strong anthropic principle," stating that the laws of nature are such as to lead finally to the evolution of human life. This way of arguing, setting a final goal as a cause for evolution, is forbidden in science, it is frankly theological in a scientific disguise. Now theology and science are different, and it would be a mistake to force theology to be a branch of physics. Therefore we shall tolerate theological principles like the "strong anthropic principle" as of metaphysical or theological value, and discuss possible clashes or concurrences with scientific reasoning.

Obviously these aspects provoke speculations of all kinds, also of theological importance. It is very suggestive to suppose a divine plan behind such a tailor-made universe. The more we investigate the connections between forces and constants of nature, the more we find evidence for a precise fine-tuning which enables life of our type to evolve. To that we may add arguments from the theory of evolution which indicate that the biological evolution toward human beings has moved along a narrow, precarious path.

But these arguments cannot serve as proofs in a scientific sense, just as the theological or cosmic proof of God's existence put forward in the Middle Ages cannot be accepted in science. Whenever a specific, actual situation is described, it becomes more and more improbable, as more of its characteristic properties are considered.

There are proponents of the anthropic principle, who understand it as a kind of selection principle. The chaotic inflationary model put forward by Andrej Linde with its multitude of bubbles, causally disjoint and perhaps even equipped with different laws of physics, different constants of nature, would be a "many-world" model which may be at least a logically acceptable possibility. Among the many bubbles there is at least one which is equipped with a combination of constants and laws of nature suitable for us. Just as you will probably find in a department store a suit that fits well, if there are many choices, there will be among the many "universes" one which allows the evolution of life. The mysterious fine-tunings are thus no longer mysterious, they are a trivial consequence of life finding the right one among the multiple-choice universes.

A number of theoretical physicists, among them the Nobel Prize laureate Steven Weinberg, find life in the "multiverse" – one of those ugly names coined by people, who apparently have no sense for classical Latin or Greek – apparently quite attractive. Apart from the inflationary model there are other speculations which also propose a multitude of worlds. If our universe has jumped out of a quantum vacuum as a random fluctuation, such processes of creation could happen again and again. A fundamental theory, like "string theory" with its complex vacuum structure may – as some speculations say – produce a variety of worlds quite naturally. Anything possible according to logic might in some sense exist in the multiverse. It appears to me that these considerations try to evade an answer to the question: "Why is our universe as it is?" I also feel that they are not very economical: To have billions of universes inflating and decaying to finally create the possibility for life on an insignificant planet at the edge of a galaxy seems a high investment. We do not even know which connections between the quantities of physics might not have been discovered yet. Therefore it seems premature for the physicists to argue for parallel universes, before the urgently sought after theory of everything has been formulated. We should not throw the towel into the ring too early.

The founder of the theory of gravity, Isaac Newton, has supposed that the fact that all planets move around the Sun in a plane is due to the will of the Creator. According to Newton's theory every planet might move in its own orbital plane, with an orientation different from the others. Today we think that the rotation of the primeval solar nebula made it collapse into a disk, and for this reason the planets move in the plane of that disk. An obvious astrophysical explanation. I expect that many at present inexplicable fine-tunings will be resolved in a similar way in the future.

The anthropic principle is of some importance for physics, because it points out relations which need to be explained. It cannot pass as a principle of physics, however. We may let it pass as a metaphysical argument, and as an indication that the world is made hospitable for us – by whomever or by whatever means.

2.7 How Will It End?

The dark energy accelerating the expansion of the universe will dominate the cosmic evolution for some time. If it is really a cosmological constant, the expansion of the universe will go on without end. It might also be the case, that the dark energy is the energy of the ground state of some field, and that it only appears to be a constant at present, but can change over large cosmic time spans. Then it depends on the time development of the field whether a new big bang will occur, or new particles will be created continuously from the stock of dark energy. Let us not dwell on these speculations now, because there is no experimental indication of a deviation of the dark energy from a constant value.

The continuing cosmic expansion in the case of a constant energy density is of decisive importance for the final state of the cosmos. When there is no end of time in the future, every physical process, even the slowest one, can run to its end.

The biosphere of the Earth will perish in 5 billion years, when the Sun will blow up to become a red giant extending beyond the Earth's orbit.

After that it will gradually get darker, because the stars will be extinguished after they have used up their nuclear fuel and the last supernova explosions fade away.

Systems bound by gravity will radiate away their energy in the form of gravitational waves according to Einstein's theory of gravity. Since 1978 we have learned from observations of the binary pulsar 1913+16 that its orbit changes exactly as the formula for the energy loss by gravitational radiation predicts. Thus over tremendously long periods of time, much longer than the actual age of the universe, all gravitationally bound systems will radiate away their energy of motion, and the bodies will crash into each other, finally ending in black holes. Gigantic black holes, each one made from all the stars in a galaxy, then move away from each other in a dark cosmos. At the same time it gets "colder" and "colder," because the temperature of the microwave background radiation keeps falling. Such is the dreary picture of the end of the universe as the cosmologists draw it.

The drawing is not yet complete, if Stephen Hawking is right with his hypothesis put forward in 1974 that black holes can evaporate. After about 10^{70} years all the black holes should have evaporated, and only very long-wavelength radiation would remain. As in the beginning the universe is filled with radiation, at the end of a temperature which is gradually approaching absolute zero.

The predictions for the end of the cosmos are extrapolations into a very distant future, and they depend on our present state of knowledge. May things turn out quite differently? Probably not, because physical processes relentlessly drive the evolution toward the final state, where everything is dead and cold, and swallowed by black holes. Only if, as we have said, the field responsible for dark energy developed in an interesting way, then quite a different story might unfold.

The picture is incomplete in yet another aspect. The fact that intelligent life has evolved in the cosmos has not been taken into consideration. How far and in what direction can a technically oriented culture evolve in a few billion years? This will be beyond our imagination. But we can speculate about the question whether the boundary conditions of the expanding universe imply that intelligent life has to come to an end. The physicist Freeman Dyson living in Princeton has indulged in such speculations. He reaches the optimistic conclusion that intelligent life can survive forever, if it learns to adapt itself to any type of environment. Even the cosmos consisting mainly of black holes would not be completely dead. Between long and quiet phases there would occur every now and then a burst of gamma radiation, when one of the black holes evaporated. During quiet phases life would be in hibernation, and become active only during short intervals to make use of the newly produced energy. There would be no end to such activities.

For mankind on the Earth, as we have mentioned already, there is an end to comfortable living, when the Sun blows up to become a red giant star. This event in about 5 billion years will destroy the Earth's biosphere. It is the task of future generations to survive that. Five billion years is a substantial time span available for the evolution of human intelligence. I do not doubt that we will learn during this time to control simple, astronomical events or to escape from them. Then mankind has a long future ahead.

2.8 Extremes of Space and Time: Big Bang and Black Holes

The universe begins with the big bang, a singular state out of which space and time, matter and radiation arise, and where there is no before. We have discussed its remarkable features in the last section, and to do that we had to look far into the past. But even at present there are celestial objects in our Milky Way which have similar properties, namely the black holes. Their existence is predicted by Einstein's theory of general relativity as an extreme state of matter: The black hole which lets neither matter nor radiation escape, but swallows everything which gets within its reach of attraction, is like a mirror image in time of the big-bang singularity from which space, time, and matter escape, but nothing is swallowed.

These singular states of a physical theory seemed so strange at first that it was a long time before physicists accepted them as real things to be taken seriously. Meanwhile these fascinating structures are mentioned like commonplace objects in movies, journals, and books of all kinds.

It is a fundamental belief of physicists that all quantities in nature are finite and exactly measurable. Singularities in physical theories are seen as the consequence of a faulty mathematical formulation, or an expression of the intrinsic incompleteness of the theory. In this sense GR theory predicts its own failure, the limits of its validity. It must be replaced by a more general theory which overcomes these limitations. Many physicists are convinced that a theory linking quantum mechanical conceptions with properties of Einstein's GRT is needed. One might think of quantizing GRT, but despite zealous efforts there has been no success. On the other hand one could imagine a more fundamental quantum theory encompassing GRT such that the classical theory would result as a well-defined approximation. String theory claims to have achieved just that, but this approach as far as it has been developed now gives us no more than a vague guess, as to how the problem of singularities might be resolved.

Intuitively it is obvious that big masses must suffer a catastrophic fate by the action of gravity, because the gravitational force is attractive for all massive particles in the same way. In addition it is long-ranged, i.e., it decreases slowly with distance (proportional to the inverse square of the distance, to be exact).

If we add more and more particles to a given mass, the gravitational force pulling the particles together grows proportionally to the number of particles. It will finally dominate over all kinds of pressure forces which might oppose its contracting power. The thermonuclear pressure in the interior of a star, the Fermi pressure of a cold gas of electrons or neutrons, and the repulsive force between nucleons very close to each other, all will be overcome by the gravitational force, if the mass of the body is sufficiently big. In addition not only matter, but also antimatter, as well as all forms of energy act gravitationally. Furthermore any kind of energy feels the pull of gravity. Therefore a huge inner pressure can balance a big mass up to a certain limit, but since the pressure also contributes to the gravitational force eventually it becomes itself responsible for the collapse.

At first these singularities were thought to be a consequence of the strong symmetry conditions which had to be imposed to find solutions of the complex equations of general relativity. The hope was that less symmetric, slightly changed solutions would not possess such singularities. In the period from 1965 to 1970 the British mathematicians and physicists Roger Penrose, Stephen Hawking, and Brendan Carter showed that singularities of space–time occur in general nonsymmetric cases (they are "generic") and are essentially stable against small perturbations.

The singularities themselves cannot be investigated without a theory of quantum gravity, but we can try to describe the space–time structure in the environment of these infinities. What happens close to black holes? The conceptions behind words like "black hole," "space–time," and "gravitational collapse" involve the intellectual power of Einstein's theory of gravitation (we often abbreviate "theory of general relativity" as "GRT"). We should make ourselves familiar with a few fundamental features of this theory, before we discuss the extreme aspects.

2.8.1 Space and Time

The conceptions of space and time are intuitively well known to everybody: Space has three dimensions, is infinite in all dimensions, and basically nothing else but the stage for all kinds of physical processes. According to the classical picture each localized event occurs at a definite time. We know from our everyday life quite well that a rendezvous can only be successful, if we agree upon both place and time.

In a train timetable, e.g., place and time of the stations along a route are given, and we can display this graphically in a space–time diagram as in Fig. 2.20.

This drawing shows the "world line" of a passenger, who travels from Munich to Stuttgart. Already this trivial example shows the possibility how to survey motions completely in a space–time diagram. More complex situations like the motion of masses in a plane or in space can be represented in the same graphic way: Kinematics is just space–time geometry. Classical mechanics established by Newton can be displayed in this graphic way which is nothing but a translation of the usual notions of space and time.

Newton assumed that space was "absolute" with geometrical properties which already Euclid had derived from a few assumptions as, e.g., the axiom of parallels. These assumptions were

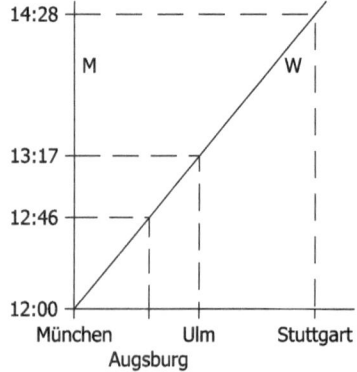

Fig. 2.20 A train timetable can be represented schematically as a space—time diagram. The graph can be interpreted as the world line of a passenger travelling from Munich to Stuttgart

considered by Newton self-evident, and he did not even mention them explicitly. In addition Newton supposed that a universal "absolute time" existed fit for all processes and measurable. His idea was that absolute space with its fixed metric was the unchanging background, where bodies moved according to a given absolute time measure. In Fig. 2.20 the axis for localities is an illustration of the absolute space, and the time axis of the absolute time of Newton. In classical, nonrelativistic physics it is simply supposed that of two events one can always determine their relative position and the time interval between them, although it is not really clear what is meant, when two events far apart from each other are said to happen simultaneously.

The motions actually occurring were explained by Newton in his famous law which states that each body, because of its inertia, is either at rest or in uniform linear motion. Forces acting on the body deflect it from this undisturbed motion.

One of these forces is gravity. Newton describes it as the mutual attraction between two massive bodies proportional to the product of the two masses and inversely proportional to the square of the distance between them.

Albert Einstein has replaced Newton's assumptions on space and time in two steps by new assumptions which are somewhat more general, and which fit the real situation even better. The desire to find a better theory developed in the wake of Michael Faraday's experiments and James Clerc Maxwell's theory of electromagnetic phenomena. It seemed that electric fields had to be considered as real physical objects. It follows from the field equations that electromagnetic waves in empty space propagate with the velocity of light. According to Maxwell the propagation does not depend on the motion of the light source, but only on the emission event. How can this be understood? Wouldn't we expect the velocity of light to be higher or lower depending on whether the source approaches us or moves away, respectively? These difficulties led Einstein in 1905 to abolish the conceptions of absolute space and absolute time, because they were not appropriate to describe such processes.

He began with two basic assumptions: The velocity of light ought to be independent of the motion of the source, as in the theory of Maxwell, and it ought to be an upper limit for the velocity of signal propagation. Both assumptions have meanwhile been confirmed many times in experiments.

The theory of special relativity (SRT) developed from this starting point has several remarkable and surprising consequences. Time no longer flows uniformly at all points of space, like Newton's absolute time, but the flow of time depends on the motion of the clock or of the observer, who by some means measures time. Moving clocks have a slower rate than clocks at rest. We do not realize this normally, because the effects are tiny, as long as the velocity of the clock is small compared to the speed of light. The slowing down of moving clocks has been demonstrated several times by comparing an atomic clock at rest on the Earth with another one transported in an airplane. The moving clock was really slower by a few billionth of a second, when both clocks came together again.

The change of the flow of time becomes distinctly noticeable, if the motion is very close to the speed of light, such as for particles in one of the big accelerators. It has been shown that particles which at rest would decay within fractions of a second, survive several seconds while racing around in the accelerator. There is also the famous "twin paradox" as another consequence of this effect:

One of the twins stays on the Earth, while the other one travels for several years with high velocity in space. When they meet again, the twin on the Earth has grown older by just so many years, while the traveler has remained young. The faster he has traveled, the younger he has stayed. How can this be reconciled with the principle of relativity? Each of the twins moves relative to the other, and it should not matter, who is considered to be at rest or to be moving. That is true, but when the twin traveling away from the Earth returns home, he has to change his direction of motion, and he must decelerate such that the relative speed between the twins becomes zero. These changes of the straight, linear motion (called "accelerations" in general by physicists) distinguish the traveling twin from the one at rest, and therefore they find a real difference in proper time, when they compare their clocks. In Fig. 2.21 we see the space–time triangle illustrating the twin paradox. At U the direction of motion of the traveler changes, and at R it again changes, when he comes to rest.

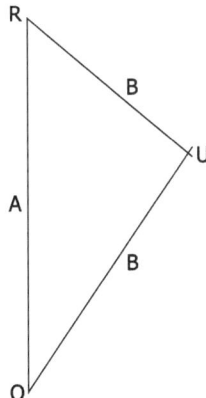

Fig. 2.21 In space–time the triangle inequality is valid in the unusual form that the distance OR (the proper time interval from O to R) is greater than the sum of the space–time distances OU plus UR. This is the so-called twin-paradox The twin at rest (path A) experiences a longer time interval between events O and R than the travelling twin, who moves from O to R along path B (Time is vertical, space horizontal in this diagram)

This experiment has not been carried out yet, but there are no doubts about the outcome, according to all we know about the space–time of SRT.

Very significant is the fact that in general we cannot decide whether spatially separated events are simultaneous or not. This depends on the state of motion of the observer. The classification of events into "earlier" or "later," the distinction between "before" and "after" is in general not unique, but different for different observers. If we see two events A and B such that B follows A, then an observer moving relative to us with the appropriate velocity might conclude that A follows B, i.e., exactly the opposite temporal ordering of events.

In a space–time diagram we would plot the world line of an observer, who moves with constant velocity compared to an observer at rest, as a straight line tilted against the vertical time axis. Events which are judged as simultaneous by the moving observer lie on a straight line which is tilted toward the world line away from the vertical direction. Moving and nonmoving observers have therefore quite a different view of the order in time of events (Fig. 2.22).

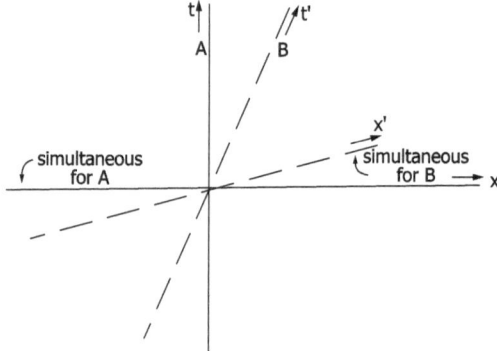

Fig. 2.22 All events considered as simultaneous by a moving observer B lie on a straight line which is tilted against the horizontal axis. The horizontal axis is the line containing all events simultaneous for an observer at rest A. Any event in the area between the two lines of simultaneity is considered by A to happen in time after the event at the origin of the coordinate system, by B to occur before

These consequences of Einstein's theory have found great public interest about a hundred years ago, when they were put forward. Probably the feeling was that these insights pointed to our redemption from the inflexible law of temporal order, of the "before" and "after" which now depended on the state of motion and had thereby lost some of its importance.

The light rays moving out from a certain space–time point form a surface in the space–time, the "light cone" of that point (Fig. 2.23). The totality of light cones has a deep significance: Our experience tells us that no signal can propagate faster than light. Therefore inside the light cone of a space–time point are all the events which can be reached by signals from the tip of the light-cone. We can also draw a "past light-cone" for each point, i.e., the surface consisting of all the light rays reaching that point. Inside the past light cone are all events which can reach the point at the tip of the light cone by signals (see Fig. 2.24).

As we can see in Fig. 2.24 this divides the environment of each space–time point into separate regions. This "causal structure" of space–time can be described mathematically by the consideration of the four-dimensional space–time distances of events: For two events the distance between the two points in space and the time interval are combined in the following way:

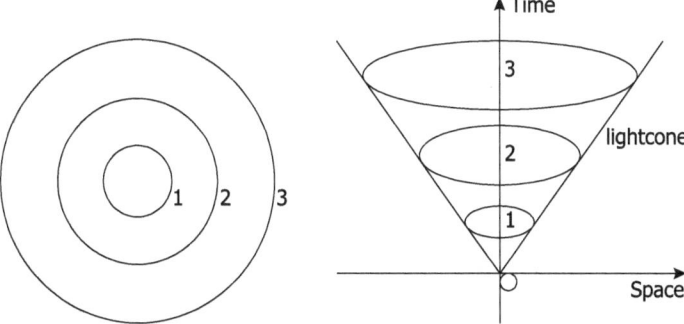

Fig. 2.23 Taking instead of 3D space a 2D surface we can graphically present the propagation of light rays. The signals of a source of light at the coordinate origin 0 reach larger and larger circles around 0 with increasing time. In a space–time diagram we can draw these circles above one another along the time axis. They form a cone with the tip at 0. The opening angle of the cone depends on the measure used for spatial distances. One can choose the unit such that this angle is a right one, i.e., 90° (this is achieved, for example, by choosing seconds as time units, and light-seconds as length units). Light rays then follow straight lines bent by 45° against the vertical time axis, and also by 45° against the horizontal space plane

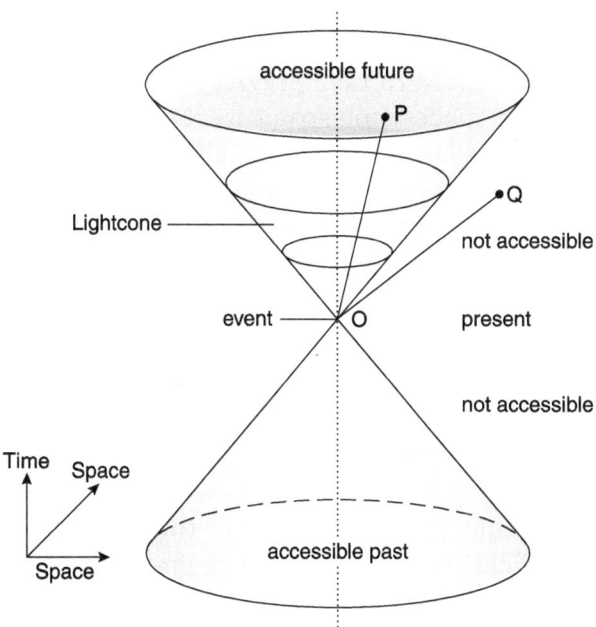

Fig. 2.24 The light cone of an event 0 is the set of events which can be reached by light signals from 0, or from which 0 can be reached

Squaring the time interval after multiplication by the velocity of light (i.e., squaring the distance a light ray would reach during that time interval), and subtracting from it the square of the spatial separation gives the square of a quantity which is commonly called "space–time distance," and designated by s. This quantity is zero along light rays, i.e., on the light cone (by construction); the square of s is positive for events inside, and negative for events outside of the light cone. A series of experiments have shown that s is the time measured by clocks in space–time.

The "twin paradox" is seen to be just the somewhat unusual inequality which holds for triangles in space–time (Fig. 2.21): The twin at rest moves along path A from O to R, and measures a proper time interval $s(OR)$. The twin along path B travels first from O to U and measures $s(OU) + s(UR)$ as his proper time interval. Now $s(OR)$ is greater than the sum $s(OU) + s(UR)$, as you can easily check. This is the opposite to the relation for a triangle OUR in Euclidean geometry.

The special theory of relativity is established firmly as the foundation of all parts of physics where gravity can be neglected. Especially for high-energy particle physics the theory is indispensable.

Einstein suggested to take gravity into account by a second refinement of the space–time structure. His theory of gravity, the theory of general relativity of 1915 (GRT), is based on the idea that the structure of space–time is not fixed, but determined by the masses and energies present. Each massive object distorts the space–time metric in its surroundings. The dynamics of the massive bodies, on the other hand, is determined by the geometry of space–time. Thus different objects act on each other: This is gravity. The metric field between the bodies acts also on light which is deflected near massive objects. Consequently the light cone structure is not fixed, but results from the distribution of the masses. This tight connection of space, time, and matter causes problems for computations in GRT. But this theory was a big step toward the unification of physics. Since gravity is woven into the geometric properties of space–time, there is no need to treat it as in Newton's theory like an additional independent structure. In that sense GRT is the most beautiful part of physics.

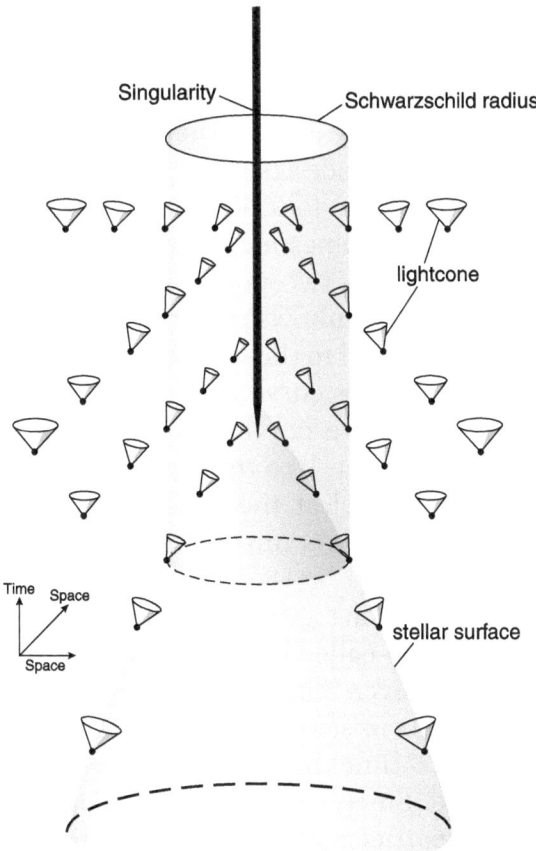

Fig. 2.25 A star collapsing to a black hole shrinks to a point, where the whole mass is concentrated with infinite density. In this space–time diagram the star is a circle contracting to zero radius. The light cones bend toward the mass. One sees that before the final singularity is formed a surface in space–time exists, where light can no longer propagate outward. The light cones at this surface all point inward toward the singularity. This surface is the Schwarzschild horizon, its radius is the Schwarzschild radius ($r_s = 2GM/c^2$; G: constant of gravity, M: mass, c: velocity of light). The Schwarzschild horizon is a null surface; its existence prohibits for any outside observer to see the black hole

In Fig. 2.25 the change of the light cones in the vicinity of a mass is shown. Close to the mass the light cones bend inward, that is the light rays are curved. This deviation from a straight line propagation is due to the "curvature of space" caused by a massive body according to GRT.

We may try to grasp this somewhat difficult idea by looking at the orbits of freely moving particles. In the four-dimensional, flat space–time, the Minkowski space, the particles move along straight lines, if no forces act on them. The straight lines in Minkowski space are "geodesics," and there are also geodesics in curved space–times. The four-dimensional distance measured along geodesic curves becomes maximal. (In Euclidean space the shortest distance between two points is given by the geodesic connecting them). As in Minkowski space the geodesic distance in curved space–times is equal to the proper time interval s measured by a clock carried along this curve.

In the case of positive curvature two "parallel" geodesics converge toward each other, like the big circles on the surface of a sphere which are parallel at the equator, and intersect at the poles. Negative curvature lets initially parallel geodesics diverge (cf. Fig. 2.12).

Light propagates along geodesics, along which any proper time interval is zero, so-called null geodesics. A photon does not experience time at all, even if it covers cosmic distances of billions of light-years between emission and reabsorption.

In a curved space–time the null geodesics form the light cone just as in a flat one. Since signals cannot propagate faster than light, the paths of particles of positive mass lie inside the light cones defined at every point of the path. Thus the light cones are the boundaries for the propagation of signals, they mark the causal structure of the space–time.

2.8.2 Gravity

Gravity by itself is no more in Einstein's theory. It is built into the space–time texture and has become part of the geometry. But how then can it determine the mutual attraction of masses and the motion of heavenly bodies?

Let us look for a moment at the solar system: According to GRT the orbit of the Earth around the Sun is determined by the distortion of the space–time in the solar neighborhood. Imagine the space–time structure like a sheet of elastic material. A big massive body like the Sun produces a deep dent. Bodies which come too close to this funnel fall down toward the Sun. This is the

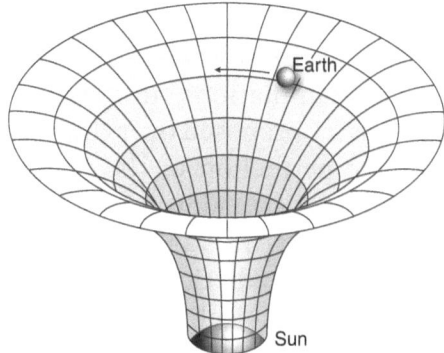

Fig. 2.26 Gravity is transformed to a geometric property of space–time in Einstein's GRT. The figure displays schematically how the Sun with its big mass is at the bottom of a deep funnel in space–time which we illustrate as an elastic rubber sheet. The Earth rolls along the funnel wall, prevented from crashing down into the Sun by its own velocity

attraction between masses which looks a bit one-sided in the case of the solar system, because the mass of the Sun is so dominant (Fig. 2.26).

The Earth can move on its path at the wall of this funnel, because its velocity is just right to keep it there – as Kepler's law requires.

The Earth has its own mass, and it forms a smaller funnel with satellites moving along its walls. Since the mass of the satellites is negligible, they do not experience any gravitational force inside, everything is weightless, at zero-G. The gravitational attraction of the Earth is compensated by the orbital motion. Practically everybody has seen TV transmissions of astronauts floating around in the space station, where the effects of weightlessness have been demonstrated impressively. These effects are examples for one fundamental principle of Einstein's theory, the "principle of equivalence." It states that at any one point the action of gravity cannot be distinguished from an appropriate acceleration. Along its ellipsoidal orbit around the Sun the Earth is exactly in this balance between acceleration and gravitational attraction.

It is astonishing that despite its very different approach GRT transforms into the Newtonian theory of gravitation, if gravitational fields are weak and velocities are small.

On several points, however, GRT corrects the predictions of Newtonian theory. Up to now GRT has successfully met all critical experimental examinations, in contrast to a number of rivaling theories of gravity proposed over the years. It explains the anomalous shift of the perihelion of the planet Mercury which has been known since 1859. In addition many observations have confirmed the prediction of the deflection of light rays by the Sun's gravitational field, a prediction which was first found true during observations of a solar eclipse in 1919. Many other and much more precise tests have meanwhile succeeded to establish the combined influence of gravity and motion on the rate of clocks, and the delay of radar signals due to the gravitational potentials they cross. All these observations and measurements constitute solid proofs for Einstein's supposition of the curvature of space–time, and the dynamical properties of the metric.

For atoms gravity is completely insignificant, but it gains importance as large masses come into play. The other known long-range force, the electromagnetic force, acts differently on positive and negative charges. Particles with the same charge repel each other, with opposite charge they attract each other. This leads to a screening of the electromagnetic force at large distances. There seems to be no screening effect for gravitation, because negative masses are not known.

2.8.3 Black Hole Basics

Two months after Albert Einstein had published his fundamental paper on GR, the brilliant German physicist Karl Schwarzschild derived at the end of the year 1915 the solution later named after him. The "Schwarzschild Geometry" has become famous as the prototype of a "black hole," although it gives a general description of the curved space–time outside of a spherically symmetric distribution of masses. The space around the Sun is also approximately described by the Schwarzschild geometry, as far as the Sun can be considered as a perfect sphere.

Let us imagine compressing a spherical mass, like the Sun, into a smaller and smaller radius. In Newtonian theory this is possible (at least in our imagination) until the idealization of a point mass with zero radius. Outside of the mass the Schwarzschild solution is valid at all times. But for radii smaller than

$$r_s = \frac{2GM}{c^2}$$

(M is the mass within r_s, G the gravitational constant, c the velocity of light.) the static coordinates, i.e., the coordinates independent of time, which Karl Schwarzschild had used lose their validity. In this thought experiment therefore we do not reach the equivalent of the Newtonian point mass concentrated at $r = 0$, but we are stopped at the "Schwarzschild radius" r_s, and we do not know what happens inside of r_s, until we find a better description. In the course of time theoreticians have explored different coordinate frames which are better suited to describe the inner part at radii less than r_s.

The easiest approach to study this space–time is to analyze the propagation of light. Far away from the Schwarzschild radius the situation is the same as in Minkowski space, but close to the Schwarzschild radius the light cones are being deformed. A light ray emitted at the Schwarzschild radius in any arbitrary direction will be bent so much that it cannot escape to the outside, but must remain caught within this sphere of radius r_s. Light rays emitted within this radius end necessarily in the singularity at $r = 0$ (Fig. 2.27).

Light from the outside can fall into the Schwarzschild radius without any problem, but no light signal can escape.

The Schwarzschild radius thus designates a structure in the space–time, a "horizon" which separates inside and outside inescapably. Since information cannot be transmitted faster than with the velocity of light, the horizon acts like a membrane letting energy and information pass only in one direction, from the outside to the inside.

Inside at $r = 0$ is the singularity, just like the Newtonian point mass. Every mass inside the horizon must necessarily end as a singular point mass according to GRT. Inside the horizon the light cones turn by 90°. It is as if space and time had changed their properties. Inside the horizon space "passes," just as does time in the real world. Nothing can stop the passing away of the distance to the point $r = 0$, everything ends up in this singularity.

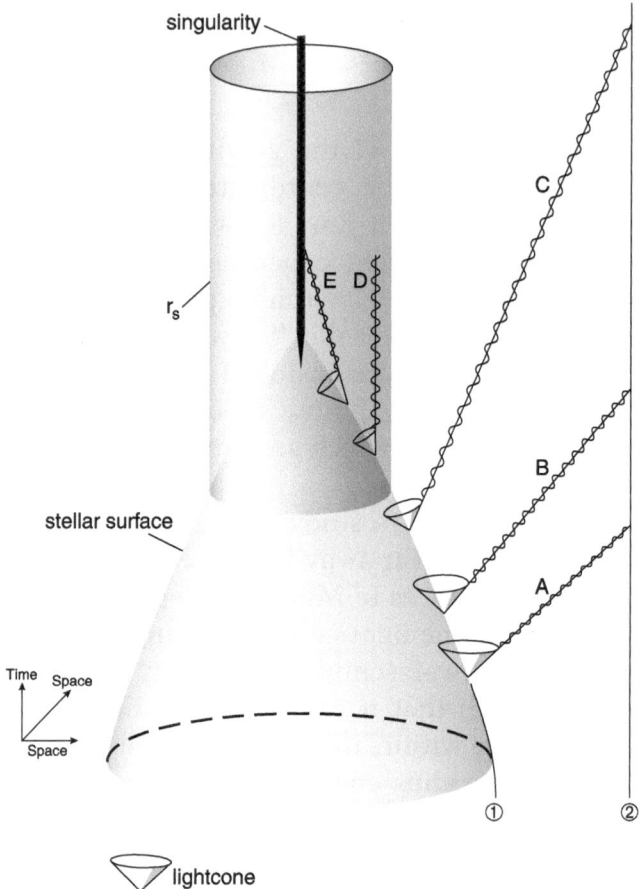

Fig. 2.27 A black hole separates space–time into two regions, inside and outside of the horizon, such that signals cannot propagate from the inside to the outside. Signals emitted at regular time intervals by a source falling into the black hole will be received by an observer outside with ever increasing delays. This time delay grows beyond any finite limit as the Schwarzschild radius is approached

The space–time of the Schwarzschild solution thus is empty, outside and inside the horizon there is no matter except for the point mass M in the center. An outside observer measures a gravitational force corresponding to a mass M inside. This structure, the point mass M surrounded by an horizon r_s is called "black hole," a name invented by the American physicist John Wheeler in a lecture in the year 1968.

(Actually there is a short novel written in 1905 by Hubert von Meyrinck with the title "The black sphere," where the emergence of this all-devouring nothingness is depicted. Meyrinck, however, ascribes the origin of the black hole to the materialization of the thoughts of an officer of the Habsburg K. and K. army.)

A black hole thus is neither a material body, nor does it consist of radiation; it is literally a hole in the space–time. The singularity inside does not have any possibility for a causal connection to an observer outside. Such horizons or causality boundaries are a remarkable result of GRT which may be characteristic for all realistic singularities: A "cosmic censor" covers it, such that no observer can see it.

Stars, planets, or other physical objects are, of course, much more extended than their Schwarzschild radius. The Schwarzschild radius for the Sun is 3 km, for the Earth 0.9 cm. The fact that Sun and Earth are much bigger, shows that the deviation from an Euclidean space is very small in the vicinity of these celestial bodies. Subatomic particles like protons and neutrons are larger than their Schwarzschild radius by a factor of about 10^{39}. This demonstrates that gravity is totally unimportant in the world of elementary particles. The situation is quite different when we look at a neutron star, an object with a mass like the Sun, but with a radius of only 10 km, about three times larger than its Schwarzschild radius.

The structure of neutron stars is definitely deviating from a purely Newtonian gravitational equilibrium. You need to squeeze a neutron star just to about one-third of its size to transform it into a black hole.

Karl Schwarzschild himself was worried by the singular behavior of his solution at the Schwarzschild radius. Therefore he investigated the properties of spheres of constant density within GRT. He was able to show that the radius of such a sphere must always be larger than $9/8\ r_s$. A sphere of this kind is always outside of the Schwarzschild radius.

Because of the enormous concentration of mass in a small volume, the tidal forces in the vicinity of a black hole are also larger than close to a normal star. Thus, a researcher falling into a solar mass black hole will be distorted by the tidal forces, outside of the horizon. Since his feet are more strongly attracted than his head, he will be stretched along the body axis, and compressed from the sides. The tensions occurring are very big – 10,000 to 100,000 g for a black hole of solar mass (g is the acceleration on the Earth's surface) – but decrease in inverse proportion to the square of the mass. The tidal forces near big black holes can be well tolerated, and the fall into the horizon proceeds quite uneventfully, even when the horizon is crossed, until it meets a horrible end at the singularity $r = 0$.

2.8.4 Gravitational Collapse

A big, massive star which has used up the nuclear fuel in its interior contracts more and more under the influence of its own gravitational field, and disappears finally in a singularity after it has crossed the horizon.

For an observer on the stellar surface this happens in the free fall time needed to cross the distance to the horizon, that is for a star of a few solar masses in a few seconds. For an observer outside, however, the stellar surface appears to come closer and closer to the horizon, but never reaches it. The star seems to "freeze" at the Schwarzschild radius. These impressions depend on the propagation of radiation in the vicinity of the horizon. Signals emitted at regular intervals by the bold researcher falling into the black hole are received by the cautious observer outside at steadily growing time intervals. Finally the last flash of light sent out at the horizon is not received by the outside observer – "it reaches him after an infinite time has passed" as relativity theorists like to phrase it (see Figs. 2.25 and 2.27).

All this is correct in principle, but in practice the star becomes invisible quite suddenly. This is due to the fact that the wavelength of the light emitted close to the horizon is received far away strongly red-shifted. The shift to longer wavelengths increases exponentially with the source of light approaching the horizon. The luminosity decays rapidly too. During the time span of one hundred thousandth of a second for a solar mass star ($T \sim r_{s/c} - 10^{-5}(M/M_\odot)$ seconds is the time needed by light to cross the Schwarzschild radius r_s), the collapsing star becomes invisible.

The Schwarzschild solution depends only on the mass, but almost all stars rotate, and therefore we expect that black holes resulting from the collapse of a star will in general rotate. Compared to the Schwarzschild solution the space–time of a rotating black hole has quite a complex horizon structure. It is also possible, at least in principle, that black holes carry an electric charge. Mass, angular momentum, and charge completely identify a black hole for the outside world. It is amazing that stars despite their rich variety of forms and structures end up in such a simple state. It reminds one of Platon's idea of "ideal bodies."

Can Black Holes Be Observed?

Black holes do not emit radiation, but they can be observed indirectly, if they accrete matter which heats up while falling into the black hole. The radiation emitted by this in-falling matter can be registered.

Black Holes in X-Ray Binary Stars

Observations of the sky in X-rays by satellites uncovered the existence of binary stars, where a star visible in optical light is orbited by an invisible, compact X-ray source. In several cases the astronomers have concluded that the X-ray source must be a black hole. This conclusion rests on estimates of the mass made possible by using the periodic fluctuations of the optical and the X-ray light to obtain accurate measurements of the system parameters. If the mass of the compact X-ray source turns out to be significantly larger than the maximal mass of a neutron star, then it must be a black hole.

The most famous candidate is the X-ray star Cyg X-1 (Cygnus X-1), with a mass between 9 and 16 solar masses, at least three times the mass limit for neutron stars. Cygnus X-1 is very probably a black hole. There are similar arguments for a number of further candidates. Astronomers would be much happier, if they did not have to rely on such indirect proofs, but were able to identify a characteristic property of the radiation – such as a typical time variability – as the unique signature of a black hole. At present such a property is not known.

Black Holes in the Centers of Galaxies

Detailed observations of the central regions of active galaxies and quasars showed evidence for high concentrations of mass and energy, for high velocities in relatively small volumes. Thus the active galaxy M87 has been searched in detail by the Hubble Space Telescope, to name but one example. Its inner region of 500 light-years contains a mass of gas which rotates with a velocity of about 750 km s^{-1}. Such a fast rotation can best be explained as the motion around a black hole of about one billion (10^9) solar masses.

Even at the center of our Milky Way a black hole has been tracked down. Measurements in the infrared part of the spectrum disclosed stellar motions in a small region of 0.3 light-years extent influenced by a central mass of about one million (10^6) solar masses. The assumption that this must be a black hole seems plausible, because any other configuration – a dense star cluster, a giant star – would not be stable, and would anyhow evolve into a black hole within a few million years.

2.8.5 Quantum Theory and Black Holes

The Singularity

The gravitational collapse of a large mass proceeds relentlessly inside the horizon until everything is concentrated in a point-like singularity. Such a singularity, in this case a point mass of infinite density, should not occur in a well-defined theory. GRT leads necessarily to such a state, as a consequence of the overwhelming gravitational force which overcomes all counteracting forces. Thus the gravitational collapse of large masses is a fascinating prediction, but marks also the limits of validity of this theory. In Newton's theory the idealization of the point mass is also used, and the gravitational field of a spherically symmetric mass distribution is described exactly as if all the mass were assembled in the central point. This then would also be a singularity, but in Newton's theory the situation is quite different: Far away the field is like that of a point mass, but close to the masses the finite density of the real distribution is relevant. In GRT, however, the point mass is real, collapse to arbitrarily small volumes is the real fate of large masses. The only redeeming feature is the formation of a horizon which shields the outside world from the singularity.

Is there a way to stop the collapse? When the matter is compressed to larger and larger densities, the classical description must become invalid at some point. Even a very massive, large star will eventually become a small quantum object. Finally we can no longer describe space–time as a classical continuum, where the separation between two points can be arbitrarily small. Space–time must finally transform into a kind of quantized structure, perhaps with a fundamental, smallest length. A theory encompassing such a unification of GRT and quantum theory is not yet in sight, although, as we have already mentioned, it has a name already: "quantum gravity." There is a lot of activity in this field of research.

Supporters of the string theory believe that this theory, or rather the fundamental, somewhat mysterious M-theory, will contain solutions which might correspond to a quantum gravity theory. The basic building blocks of this theory are "strings" or membranes, subatomic small pieces of strings or surfaces, whose vibrations create the world from the vacuum. These vibrations take place in spaces of ten dimensions at least, but in the real four-dimensional space–time six of these dimensions are wound up somehow in tiny manifolds, such that they are not perceived. An illustrative picture might be a piece of straw which from afar looks like a piece of a line, a one-dimensional structure. Closer inspection clearly shows that it is tube-like, a two-dimensional cylindrical surface, if the thickness of the walls is neglected. Looking even closer we recognize the thickness of the walls, that is the complete three-dimensional structure.

Very similar – according to string theory – is the way in which the wound-up extra dimensions show up, when a singularity is approached. Thus, during gravitational collapse for example, the singularity does not occur. Instead the ten-dimensional vibrating string unfolds. It might be like that, and right now we cannot say more about it. The M-theory has obviously an enormous number of solutions, and up to now the ones describing our real world have not been identified.

Hawking Radiation

Instead of waiting for the theory of everything one can try to look for connections between quantum and gravitational physics by investigating quantum fields in a fixed space–time. In 1974 Stephen Hawking found an exciting result: Black holes are not really black, but they emit radiation just as if there were a body with a certain temperature at the Schwarzschild radius.

He examined the behavior of the vacuum state, the state without particles, in the space–time of a black hole. The quantum mechanical vacuum is not a quiet, empty thing, but rather full of activity with pairs of particles and antiparticles being continuously created and annihilated. Now, close to the Schwarzschild horizon there is the possibility that virtual antiparticles pass through this one-way membrane, and disappear. The corresponding particles are left behind, and gain a real existence. A distant observer registers these escaping particles as a thermal radiation.

The temperature is very low for black holes of stellar mass

$$T \sim 10^{-7}(M/M_\odot)^{-1} Kelvin.$$

But small black holes, if they existed, could reach very high temperatures, and end in a burst of gamma rays. A spectacular event which might even be observable.

By the emission of such a thermal radiation, the so-called Hawking radiation, the black hole loses energy, its mass becomes smaller. The total mass would be radiated away in a time t, with

$$t = 10^{71}(M/M_\odot)^3 sec.$$

This is for solar mass black holes many orders of magnitude larger than the age of the universe.

What remains at the end? Only radiation, or some kind of scar in the space–time texture? We do not know, because the real problem, to compute the feedback of the energy loss on the black hole space–time, has not been solved.

2.8.6 Space and Time Arise and Decay

During this short walk through cosmology and astrophysics we have met many curious and remarkable phenomena. In my opinion special attention should be paid to the insight that our commonsense conceptions of space and time are shaken up: Space is not simply fixed and eternal, time is not infinite and uniformly flowing.

In the vicinity of a black hole a part of the space–time is sealed off. The Schwarzschild radius encircles a region which can no longer have contact with the space–time outside. The fall into a black hole ends in a final crunch at the singularity, and time ends there too. Viewed from the outside all clocks falling toward the black hole stop ticking at the Schwarzschild radius – it seems, as if time did not continue there.

The reverse process seems to take place at the big bang. Nothing can fall into this singularity, everything comes out of it. Space and time also have this origin in the initial singularity about 14 billion years ago.

The German philosopher Immanuel Kant has taught us that space and time are categories of our experience, they determine the principal way according to which we order our experiences. We cannot imagine otherwise, than to order them in space and time. Kant obviously had in mind the Newtonian conceptions of absolute space and uniformly flowing, eternal time. In his time these ideas were considered to be self-evident. Modern science has brought a great change. Although it is still true that we order our knowledge in the categories of space and time, the categories themselves have lost their absolute validity. If space–time itself has been created 14 billion years ago, it cannot be an eternally durable category. Thus the categories of our mind are not something given, fixed, and absolute, but they have arisen during the long evolution of the cosmos. These insights gained from physics contradict, as it seems, our intuition, and go beyond our everyday experience: If space and time arise and perish, then there might even be structures beyond space and time. This indication of a transcendent element in the world around us is like a confirmation of Kant. There may be things beyond space and time, but they are not accessible by our experience.

It is difficult for us to believe that there may be an aspect of reality not bound up with time. Our life is dominated by time: Beautiful and awful experiences passing, the before and after, the course of days and years, aging and death are inescapable reality for us.

The changes in our view of space and time do not imply with certainty that our existence reaches beyond time and that there exists really something not encased by space and time. But we can see clearly that there are restrictions imposed on our knowledge, and therefore we can also see the possibility to remove these boundaries in our thoughts.

Thus with the fall of time from its absolute throne, with the fall of the tyrant, who determines all things, and its translation to a quantity which itself must experience changes, the hope grows that – to use the words of the Austrian physicist Erwin Schrödinger "the whole time-table is not meant as seriously, as it appears at first sight" ("dass der ganze Zeitplan doch nicht so ernst gemeint ist, wie es zunächst scheinen mag").

3. The Deep Underground: Quantum World and Elementary Particles

Properties of the cosmos such as space and time can be linked quite directly to our common everyday experience. This becomes more difficult in the realm of quantum objects, where the interpretation of experiments often seems to contradict common sense.

We have to try and grasp some basic features of the quantum world, because atoms and elementary particles are the basic structures on which the whole creation rests. The fundamental properties of these smallest building blocks determine especially the early epochs of the universe, and thus knowledge of their behavior helps our insight into the workings of the universe. The creation as a whole would be incomplete without the quantum world which after all drives the entire structure. In addition we shall see how remarkably different the quantum objects are compared to the familiar macroscopic ones. An appropriate presentation seems only possible through the language of mathematics with its high degree of abstraction. Often we do not find this satisfactory, since we have the desire to paint an illustrative picture in our mind on "what holds the world together deep inside."

The problems one meets, when one tries to translate concepts from mathematics and exact sciences into daily language, are illustrated nicely in a fictitious conversation put on paper by Hans Magnus Enzensberger in his article "drawbridge out of order":

Mathematician: It is about one of the most important discoveries of the last century.

Layman: Can you explain this in words such that a normal mortal can understand it?

G. Börner, *The Wondrous Universe*, Astronomers' Universe,
DOI 10.1007/978-3-642-20104-2_3,
© Springer-Verlag Berlin Heidelberg 2011

Mathematician:	Impossible. You cannot even have an impression of it, if you do not understand the technical details.
	How can I talk about manifolds without mentioning that the theorems concerned can only hold, if those manifolds are finite dimensional, paracompact and hausdorffian, and if they have an empty boundary?
Layman:	Well, just lie a little bit.
Mathematician:	That is not in my department.
Layman:	Why not? After all, everybody else is lying too.
Mathematician:	But I have to stick to the truth.
Layman:	Sure. But could you not bend it a little bit, if by that you can help people understand what you are actually pursuing?
Mathematician:	Well, I just might try it.

Let us risk in what follows an attempt of this kind – although we will not really lie, truth will nevertheless be bent a tiny bit here and there. As long as I am myself aware of that, I will always clearly spell it out.

3.1 The Fundamental Building Blocks of Matter

Normal matter, the chemical elements which are the constituents of the Sun, the Earth, and the planets – as well as of ourselves – account for only 5% of the cosmic substance. Nevertheless, physicists believe that these hold the key to understanding the world, since the laws deeply anchored in fundamental theories governing the behavior of the elementary particles determine also the processes involving the remaining 95% of the cosmos, namely dark matter and dark energy. That is what we hope for, and perhaps it is also true. How far fundamental insights have gone up to now shall be sketched in the following sections. A very simple question shall be posed right at the start: Can we divide up a chunk of matter again and again without end, or does the division eventually stop at smallest indivisible units, the elementary particles?

We know that molecules are not elementary, because they can be taken apart into atoms by heating and irradiation. Atoms on the other hand can be split into their components, the electrons and the nucleus, by collisions with other atoms or by irradiation with light of short wavelength. Even atomic nuclei are not elementary: They can be divided up into protons and neutrons by collisions with high-energy particles or high-energy radiation, so-called gamma rays. For about 50 years the proton and the neutron were considered to be elementary particles, but during the past 40 years high-energy physicists found that they are probably bound states of point-like particles, the "quarks." Up to now, however, proton and neutron could not be broken up into individual quarks.

It could be shown in experiments that the collisions of protons with very high energies produce a large number of particles in each collision, among them even particles with masses much bigger than the proton mass. The appearance of particles with masses larger than the mass of the incident particles is explained by Einstein's famous law of the equality of mass and energy, according to which collision energy is transformed into mass. (The formula $E = mc^2$ means that a mass m can be transformed into an energy mc^2, and an energy E corresponds to a mass E/c^2. Here c is the speed of light.)

Thus the protons participating in the collisions are not split up into still smaller units, but a whole bunch of particles with different masses is created (Figs. 3.1 and 3.2).

This could be the end of the division, but it could also be that the energies have not yet been reached which would

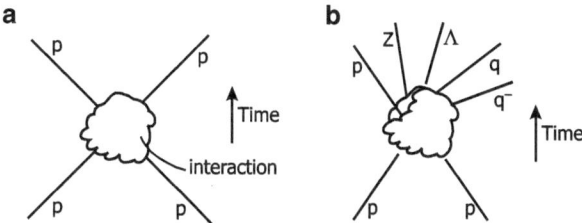

Fig. 3.1 The schematic diagram illustrates two possibilities for the outcome of a collision of two protons: Either two protons or a multitude of particles come out of the interaction region

Fig. 3.2 A collision event at the Large Hadron Collider at CERN registered with the ATLAS experiment (*courtesy of CERN Doc. Service*)

lead to a further splitting up. The final decision must be left to experiments.

To be sure, according to present-day theoretical understanding it is not the elementary particles which are the fundamental objects, but rather the fields corresponding to them. The volatile and changeable material particles are seen as excitations of the fields. We will enter into further discussions of this difficult concept below.

At the moment it seems that a so-called standard model containing a small number of elementary particles can explain all experiments quite well. We want to take a closer look at this standard model. It is also of importance for ourselves, because it describes the material basis of the elements, that is of the atoms, and therefore also the basis of our material existence.

As a first step it seems definitely useful to make ourselves acquainted with the dimensions we are talking about. An atom has a diameter of 10^{-8} cm, that is ten billion atoms laid out along each other in a line give 1 m. These and even smaller dimensions cannot easily be grasped in our imagination, but we can try to become more familiar with them by a thought experiment. Let us imagine that we had a piece of paper, say of A4 format, which can be folded arbitrarily often. If we fold it once, it is only half its size, say reduced from 20 to 10 cm. After five folds it is smaller than 1 cm, and after 25 folds our sheet is only the size of an atom. If we continue folding it, we reach after 108 folds the dimension of the Planck length, the fundamental unit of 10^{-33} cm, the realm of quantum gravity, and the theory of superstrings.

Ten atoms laid out along each other give a stretch of 1 nm (10^{-9} m). This cannot be seen with the naked eye, but the tunnel scanning microscope makes structures on this scale visible. The "nanophysicists" know how to handle such dimensions, and produce small tubes and spheres, that is, rather let them arise by stimulating the self-organization of assemblies of a few atoms.

Almost everything in the atom is empty space anyway. The electron shell makes the atom appear large. But the atomic nucleus formed by protons and neutrons which carries the mass has a tiny extent of only one hundred thousandth (10^{-13}) cm of the atomic diameter. This tiny volume contains the nuclear particles, protons and neutrons densely packed. An attractive force between the nucleons tightly binds the nuclear package. This "strong interaction" is strong enough to overcome the electric repulsion between the positively charged protons, but it is just not strong enough to bind two protons. In that case the electric repulsion still dominates. Only when one or more neutrons are added, the strong interactions win out. Already the atomic nucleus of the deuterium consisting of one proton and one neutron is a stable bound state of the strong interaction. For the evolution of the cosmos it is a big advantage that the double proton does not exist, because otherwise all the hydrogen would have ended up in double-proton nuclei in the early universe. This would have been a disaster, because then neither stars nor heavier elements would have been built.

Let us briefly recapitulate the properties of nucleons and simple atoms. Protons carry one unit of positive electric charge, neutrons are electrically neutral. A nucleon has about 2,000 times the mass of an electron. The hydrogen atom has one proton as its nucleus, and one electron in the electron shell, deuterium has one proton and one neutron in the nucleus, and one electron in the shell, the atom of helium has two protons and two neutrons in the nucleus, and two electrons in the shell.

Subatomic structure is investigated by the very direct and simple principle of shooting particles of high energy at a matter probe, the higher the energy, the better. Already in the first decade of the last century the New Zealand physicist Ernest Rutherford found that atoms consist of a nucleus and an electron shell as an outcome of his experiments. He let helium nuclei run into a thin foil of gold, and found that sometimes the helium atom would be deflected by a large angle, just as if it had collided with a tiny, concentrated mass. In the last decades elementary particle physicists discovered by collision experiments that even the proton and the neutron have an inner structure. They are both made up of three point-like concentrations of mass, the "quarks." Quarks, and electrons as well, appear in all experiments as point particles without any inner structure. One can conclude that these particles must be smaller than 10^{-16} cm. The physics of elementary particles happens on scales of 10^{-13} cm or less.

Small scales can be reached, if the momentum of the incoming particle is high. High energies are realized in huge particle accelerators which are among the most expensive research instruments and the most remarkable technical achievements of our time. These machines accelerate the particles in strong electric fields and guide them by intricate arrangements of magnetic fields on precisely determined circular orbits. The newest of its kind, the Large Hadron Collider (LHC) in Geneva which just went into operation recently, accelerates particles running in an underground circle of 27 km radius up to almost the speed of light and to energies of 1,000 GeV. (The mechanism of acceleration suggests a convenient unit of mass and energy, the electron volt.) One electron volt (eV) is the energy which an electron or proton gains, when it passes through an electric potential difference of 1 V. In elementary particle physics much higher energies are common, and thus one uses units such as GeV (giga eV, 1 GeV = 10^9 eV) or TeV (tera eV, 1 TeV = 10^{12} eV). Because of the relativistic formula $E = mc^2$ the masses of the elementary particles are often not given in grams, but in equivalent energy units. The mass of a proton is about 1 GeV in energy units. The heaviest know particle is the Z^0 with a mass corresponding to an energy of about 100 GeV ~ 0.1 TeV.

3.2 Reactions Between Elementary Particles

The large particle accelerators have been built to study collisions between various types of particles. Possible collision processes between two protons are shown in Figs. 3.1 and 3.2. Two protons collide and produce close to each other a region containing a high concentration of mass and energy. This region is unstable, and can decay in many ways into particles. In Fig. 3.1 two of the many possibilities are shown: The end products can be again two protons, or a big number of other heavy particles.

Quite often we know enough about the fundamental interactions to dissect the cloudy region of energy concentration further. Thus the reaction determined by the electromagnetic interaction, where an electron e^- and a positron (e^+ – the antiparticle of the electron with equal mass and positive charge) collide, and produce a negatively charged μ^- meson (μ^-: an elementary particle very similar to the electron, but with a bigger mass, see Table 3.1) and its positively charged antiparticle μ^+, can be drawn as in Fig. 3.3:

The collision combines the electron and the positron into an intermediate state of very high energy which rapidly decays again into μ^- and μ^+. The intermediate state has many properties of a photon, but its mass is not zero. One speaks of a "virtual photon" which does not exist as a real particle, but which appears in the computation of collision processes as a bearer or "messenger particle" of the interaction between e^+, e^- and μ^+, μ^-. The decay product of the virtual photon may also be a quark–antiquark pair, or another electron–positron pair.

Another example is the scattering of e^- and μ^+ shown in Fig. 3.4.

Table 3.1 The leptons

Particle	Charge (in units e)	Mass (mc^2)	Lifetime	Bose(B) Fermi(F)
Electron (e)	-1	0.51 MeV	Stable	F
Electron-neutrino (ν_e)	0	($\leq$)17 eV	Stable	F
Muon (μ)	-1	105.7 MeV	2.2×10^{-6} s	F
Muon-neutrino (ν_μ)	0	≤ 0.27 MeV	Stable	F
Tau-lepton (τ)	-1	1,785 MeV	3×10^{-13} s	F
Tau-neutrino (ν_τ)	0	<35 MeV	Unknown	F

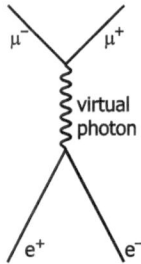

Fig. 3.3 This schematic representation of the collision of an electron and a positron shows the process in which both particles change into μ-mesons

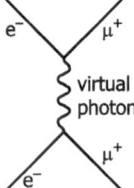

Fig. 3.4 A virtual photon carries the interaction of electron and μ-meson

These diagrams not only give an illustration of the processes, they actually define exact rules to compute these reactions. They were invented by the American physicist Richard Feynman, and are named after him ("Feynman diagrams").

Almost all elementary particles are unstable, that is, they decay in a very short time (within 10^{-22} and 10^{-6} s) into particles of smaller mass. Only a few stable particles are known such as the electron, the proton, or the neutrinos. Theories have been formulated, where even the proton may not be stable. It must, of course, have a very long lifetime, since it is one of the basic building blocks of our world.

3.2.1 Quantum Field Theory

The description of all these processes, where particles are created or annihilated with a huge energy output, relies on a fundamental quantity, the "quantized field."

Electric or magnetic fields are well known to us from everyday experience, we make use of them without paying much attention to their quite remarkable and uncommon properties.

Ever since Michael Faraday made the field lines of a magnet visible by sprinkling iron file chips on a piece of paper covering the magnet, we imagine fields like that: We imagine an electric charge or a magnet surrounded by lines which pervade empty space, and which tell us how electric or magnetic forces act. Electromagnetic fields changing with time propagate in space with the velocity of light – examples are radio waves and light waves. They can be used to emit and receive signals, and they do not need a material medium for their propagation, such as waves in water do. Electromagnetic fields cross empty space, and therefore we must ascribe to them an existence as a real object, just as we do it for material particles.

This conception of a field has been extended in elementary particle physics to include "quantized fields," and to view particles as quantum states of "fields." These fields are required to satisfy equations of motion and certain quantization rules. In contrast to the classical field, the quantum field is not considered as a real, observable object, but as a mathematical tool for the efficient description of possible particle states.

Observable properties unfold, when quantum fields act on certain states, such as the vacuum, the state of lowest energy, or one-particle, two-particle states, etc. There is no doubt that the quantized field approach is a stunning success in the case of the electromagnetic field. The computations carried out with high precision in "quantum electrodynamics" yield results which agree with experimental measurements to the last digit – for example for the anomalous magnetic moment of the electron. The quanta of the electromagnetic field are the photons, the particles of radiation, whose existence has been demonstrated by the photokinetic effect, and many other experiments.

The mathematical formalism of the quantum field allows us to describe the creation and annihilation of particles in a consistent way. A fundamental element of that description is the postulate that a state without particles, a vacuum state, exists. The vacuum is commonly also the state of lowest energy of the field.

In many states the field shows the properties of a particle. There are additional states, however, which have all the properties like mass, charge, spin of a corresponding particle, but do not satisfy the condition $E = \sqrt{k^2 + m^2}$ between energy E, mass m, and momentum k which a real particle obeys. Physicists call such states "virtual" particles. Thus the force between particles is often represented as the exchange of certain virtual particles. The vacuum of a quantum field theory, the state empty of real particles, becomes in this description a highly structured medium filled with virtual particles.

The action of electric charges "polarizes" the vacuum, i.e., the virtual particles arrange themselves such that the observed electric charge is somewhat weakened. A small shift in the atomic energy levels of hydrogen (the so-called Lamb-shift) is due to this effect. The Lamb shift has been measured as a fine structure in the line spectrum of the atom. The theoretical construction of quantum field and vacuum definitely has some roots in reality.

In a state with very many particles the quantum field changes into a classical field, as e.g., the electromagnetic field in a state with many photons. Light or radio waves are such states of the electromagnetic field.

3.2.2 The Fundamental Forces

Four different interactions determine the appearance of matter according to our present knowledge: the electromagnetic, strong, and weak interaction, and gravitation. These interactions appear in a remarkable hierarchical ordering: The size of the atoms, e.g., the hydrogen atom of 10^{-8} cm size, is determined by the electromagnetic force between the positively charged protons in the atomic nucleus, and the negatively charged electrons in the shell. This determination of size is universal, i.e., all hydrogen atoms are of equal size in their most stable state. The dimension of the nuclear particles is determined by the strong interaction which rules between their fundamental constituents, the quarks. It fixes the radius of the nucleus to about 10^{-13} cm. On the scale of the atomic nuclei the electromagnetic interaction is negligible, while the strong interaction is only effective on this and on smaller scales, and without importance for the structure of the atom. The weak interaction appears as a correction to these forces in processes such as radioactive decays, and in all reactions involving neutrinos.

Up to the dimension of solid macroscopic bodies these interactions and the forces derived from them determine all structures. On these scales the gravitational force is unnoticeably weak. Although there is also a gravitational attraction between the proton and the electron in the hydrogen atom, this force is weaker by the tiny factor 10^{-39} than the electromagnetic force. But gravitation acts as an attractive force between all particles, and it decreases slowly with distance, whereas strong and weak interactions are confined to nuclear dimensions. The electromagnetic force is screened off on average, because electric charges of opposite sign attract each other, while charges of the same sign repel each other. Thus electric charges are surrounded preferentially by partners with the opposite electric charge which weakens or annihilates the original field. Therefore macroscopic bodies are electric neutral. In the end the structure and motion of the heavenly bodies and of the universe is controlled by the gravitational force which in laboratory experiments is negligibly small.

During the last few decades particle theorists have had some success in their attempts to understand this hierarchy of interactions. The suggestion to derive both the weak and the electromagnetic interactions from one common cause was especially successful, also in experiments. According to these ideas the observed difference of the forces should disappear at sufficiently high energies, i.e., on small scales. Electromagnetic and weak interactions are thought to merge into each other above this unification energy into one unique force which at high energies determines the structures. The existence of messenger particles which transmit the "electroweak" force follows from this model, and their properties are determined. Real elementary particles which correspond to these virtual messenger particles were in fact discovered in an experiment at the European Center for Nuclear Research, CERN, in Geneva in 1983. The W and Z particles, as they are called, have exactly the masses expected from the theoretical model (83 GeV, that is 83 times the proton mass for the W, 94 GeV for the Z particle). Such a convincing experimental proof of existence has, of course, dramatically increased confidence in such model-building.

In a similar scheme the strong interaction was successfully included, and finally a lot of effort has been spent trying to understand gravitation as an interaction which stands on equal footing with the other interactions at extremely high energies. Despite these efforts ongoing now for over 50 years, the scientists did not come up with a theory combining in a consistent way quantum physics with its deep-reaching grasp of the properties of matter and Einstein's general relativity theory with its elegant explanation of gravity. The problem to find the missing link between GRT and quantum theory is the main obstacle on the way to a unified theory of all interactions.

3.2.3 Elementary Particles

All observations in elementary particle physics so far can be understood within the scheme of the four fundamental forces, and a small number of elementary particles which can be subdivided into three classes.

Particles Transmitting the Force

Forces between elementary particles can be understood as the exchange of virtual particles. Real particles corresponding to these exchange states have been identified, as e.g., for the electro-magnetic force which is carried over by virtual photons. The weak interaction is transmitted by the virtual W^+, W^-, and Z^0 particles (W^+ carries one unit of positive electric charge, W^- is negatively charged, Z^0 is neutral.) The forces between the quarks are transferred by massless particles, called gluons; these have not been discovered as real particles in experiments. Gravitation should also be carried by a hypothetical messenger particle, the "graviton," but this idea is not yet a clearly defined concept, since a quantum theory of nonlinear gravitation has not been found yet. Because of the weakness of the interaction, an experimental detection of the graviton is very unlikely.

Leptons

Leptons have a small mass like the electron. They have gravitational, electromagnetic, and weak interactions, but are not affected by the strong nuclear forces. According to our present understanding there exist six leptons, three pairs of an uncharged and a charged particle each. As far as we know there is a conservation law for the number of leptons participating in a reaction. The total number of all leptons in the final state is equal to the number of leptons initially. In Table 3.1 the six known leptons are listed.

Elementary particles like the leptons in Table 3.1 are characterized by certain unchangeable properties, such as mass and electric charge. An important property of leptons is their intrinsic angular momentum, the "spin." We can picture this like the permanent intrinsic rotation of a small sphere but with some peculiarities. While a macroscopic body can carry out any arbitrary rotation, the spin is quantized, i.e., there are only multiples of half a basic unit (this unit is Planck's constant $\hbar$). Particles with spin $0, 1, 2, \ldots$ are called "bosons," particles with spin $1/2, 3/2, \ldots$ "fermions." A macroscopic body and a boson as well appear in their original configuration after a rotation about an angle of $360°$, but the spin-1/2 particles – quarks and electrons have that property – are in a different quantum mechanical state after rotating by $360°$. They are back to the original configuration only after a rotation by $720°$.

Fermions are different from bosons also in their statistical properties. Fermions obey the Pauli exclusion principle which says that two particles cannot occupy the same quantum state. This principle determines the structure of the atomic electron shells, and the periodic system of the elements in chemistry. We know the exclusion principle from the theater or the opera: Every person needs a seat, and when the best seats are taken, you have to accept a worse one. Proton and neutron are Fermi particles and this guarantees the stability of our world. For a system of fermions in a finite volume there is a lower bound of the energy. Thus there is no fall into bottomless depths to deeper and deeper energy levels, and the world remains in existence. For bosons such a lower bound for the energy does not exist.

In Table 3.1 there are leptons listed which do not take part in the build-up of normal matter, namely the neutrinos. There is an amusing anecdote connected with their discovery in the last century. In measurements of the so-called beta decay, i.e., the reaction, where an atom emits an electron and changes to another element, it was found that the laws of energy and momentum conservation were violated. Was it possible that these fundamental laws did not hold in these reactions? Wolfgang Pauli, Swiss physicist, Nobel prize winner, and leading theorist at that time, proposed that a new particle which could not be registered in this experiment might balance the momenta. In a letter to his colleagues with the opening words "Dear radioactive Ladies and Gentlemen" he suggested the name neutrino for this particle. The neutrino should not carry an electric charge, have at most a tiny mass, and should react extremely weakly with other particles. Pauli added in his letter that he would take any bet that this particle would never be found. This shows that he was not a betting man, since there is no chance to win such a bet.

Nowadays there are many experiments involving neutrinos, and physicists have succeeded in measuring in great detail the flow of neutrinos from the Sun.

3.2.4 The Quarks

Individual, free quarks have not been discovered up to now. Physicists are more or less sure that they are always bound in more complex, composite particles like the protons. Their existence has been demonstrated only indirectly by the scattering of electrons on nucleons. One could in that way identify six different quarks. Like leptons quarks can be combined to pairs. "up" (u) and "down" (d) quark together with the electron (e^-) and the neutrino (ν_e) can be put into one "fermion family." We know from the decay of the Z^0 particle that there are only three such families. The leptons μ^- and ν_μ belong to the "charm (c)" and "strange (s)" quark, τ^- and ν_τ to "bottom (b)" and "top (t)." Each quark moreover appears in three different types which are distinguished by a quantum number called "color charge." Therefore there are 3d, 3u quarks, and so on. The quarks carry an electric charge of 2/3 or 1/3 of the elementary unit designated as e.

Table 3.2 The quarks

Particle	Charge	Mass	Bose/Fermi
Up (u)	$+\frac{2}{3}$	~0.005 GeV	F
Down (d)	$-\frac{1}{3}$	~0.01 GeV	F
Charm (c)	$+\frac{2}{3}$	~1.5 GeV	F
Strange (s)	$-\frac{1}{3}$	~0.2 GeV	F
Top (t)	$+\frac{2}{3}$	~180 GeV	F
Bottom (b)	$-\frac{1}{3}$	~4 GeV	F

Table 3.3 The hadrons

Name	Mass (GeV)	Lifetime	Composition
Proton (p)	0.938	Stable	uud
Antiproton ($\bar{p}$)	0.938	Stable	$\bar{u}$ud
Neutron (n)	0.940	880 s	udd
Pi-Meson (π^+)	0.140	2.6×10^{-8} s	$\bar{u}$d
K-Meson (K^+)	0.494	1.24×10^{-8} s	u$\bar{s}$
J or Psi: ($J\psi$)	3.097	$\Gamma = (68 \pm 10)$ KeV	c$\bar{c}$
Ypsilon (Y)	9.460	$\Gamma = (52 \pm 2)$ KeV	b$\bar{b}$

Γ: Halfwidth of the resonance

In Table 3.2 some facts about the quarks have been collected.

Any process between elementary particles can also occur with the charges of all the particles of the opposite sign, and a reflection on one point in space carried out (as x to $-x$ in one dimension). This combination of two transformations is called CP for short. It leads from particles to antiparticles. The color charges of the quarks are also transformed into the anticolor charges. Thus every elementary particle has its antiparticle partner. The subnuclear particles composed of quarks and antiquarks are called hadrons. In Table 3.3 a few hadrons are listed with their most important properties. Antiparticles are designated by a cross-bar. $\bar{u}$ for instance stands for the anti-up quark.

3.3 Symmetries and Conservation Laws

Elementary particles change in collisions and decays. To bring order into this changeable world of the elementary particles, the physicists look for properties of matter which do not change. A simple example of an unchanging quantity is the total energy of all the partners in a collision. It does not matter how the process happens, the total energy is conserved, provided the masses will be added in as part of the total energy according to $E = mc^2$. Another example of a conservation law is the unchanging total electric charge in interactions between elementary particles.

The experimental fact that individual leptons or baryons (heavy particles participating in the strong interaction like protons or neutrons) are not produced or annihilated, but only pairs of particle and antiparticle, is traced back to a conservation law for the total number of baryons and the total number of leptons. These conservation laws are expressed by the introduction of additional quantum numbers: A lepton number $L = +1$ is assigned to each lepton, a negative quantum number $L = -1$ to each antilepton. Particles which are not leptons have $L = 0$. The total lepton number in a process is the sum of the lepton numbers of all the participating particles. The total number does not change in reactions; even the production of a lepton–antilepton pair does not change L, since the total change by the pair produced is $(+1) + (-1) = 0$. The baryons are assigned a baryon number $B = +1$, and antibaryons $B = -1$. Leptons have $B = 0$, and quarks $B = 1/3$. The force-transmitting particles like photon (γ), Z^0 have $B = 0, L = 0$; their number is not conserved, and they can be produced or annihilated as single particles.

Besides conservation laws there are various symmetry principles which express very clearly unchanging properties and order in the world of elementary particles. Just like the symmetry properties of color patterns or of macroscopic bodies, the symmetries of elementary particles can be seen as the invariance of the configuration against certain transformations. Very simple examples of such symmetries are rotations in space or translations in space or in time.

The invariance against translations in time implies that only time intervals are of importance, not absolute time. An immediate consequence of that symmetry is the conservation of the total energy of all particles involved in a process.

Besides these continuous transformations there are discrete or noncontinuous transformations, and corresponding discrete symmetries. From daily experience we are well acquainted with the mirror symmetry, the change of right and left, which many objects around us show. Physicists call this symmetry P. A rotating body appears in the mirror again as rotating, but in the opposite sense. The neutrinos change under the mirror transformation into their antiparticles. Actually it is the CP transformation that changes particles to antiparticles, but neutrinos have no charge. There are few reactions which do not conserve CP. One of them is the decay of the neutral K^0-mesons.

A fundamental insight is the general invariance of all reactions under a combination of CP and time reversal T. To each process there exists the corresponding one, where time runs backward, and particle and antiparticle are interchanged – with the same results. Since almost all processes are CP invariant, the world of the elementary particles is also invariant against time reversal, except for a few decay processes like those of the K^0 mesons just mentioned.

3.4 Unification of Interactions

Electricity and magnetism as they manifest themselves for instance in a spark discharge or the orientation of the needle of a compass apparently are completely different phenomena. Already in the nineteenth century Michael Faraday and his successors carried out a series of cunning experiments which demonstrated that these are only two different faces of the same basic interaction. James Maxwell could formulate a unified theory of electromagnetism in 1862 making use of these experimental results. The propagation of light and radio waves, as well as the way electromagnetic fields act on matter, is described in this theory by simple equations. Maxwell's theory is also the starting point for quantum electrodynamics (QED). QED has been a very successful theory, giving computational results of extraordinary precision which agree with the same precision with experiments.

Therefore QED has often been used as a model for the formulation of theories of other interactions.

For the weak and strong interactions these attempts were successful. With that a certain unification of elementary particle theory has been achieved, up to a so-called standard model. A very pretty success of this approach was the clarification of the so-called renormalizability of the fundamental interactions. The equations of quantum field theories like QED cannot be solved exactly, and one has to find approximation schemes. These computations give in any quantum field theory, also in QED, divergent results, that is infinitely large numbers, for example for the mass or charge of the electron.

These physical quantities are also measured experimentally. The corresponding divergencies can therefore be renormalized, if the finite measured values of the mass and charge of the electron are inserted in place of the divergent theoretical values. The physicist can thereby utilize a very convenient feature of QED: Only a few divergent quantities must be replaced in this way to obtain a mathematically well-defined approximative theory which yields finite results for all possible processes. QED is a shining example for a renormalizable theory. For a long time, however, attempts to extend this scheme to the weak and strong interactions were not successful. The new formulation of the theories according to the structure of QED as so-called gauge theories allowed for the messenger particles which transmit the interactions only mass-zero quanta.

But short-range interactions like the weak interaction with a range of less than 10^{-16} cm require messenger particles with large masses. Thus the question had to be solved, how the fields which are required to have only mass-zero quanta in the gauge theories of the weak and strong interactions can still describe particles with a large mass. The idea of "spontaneous symmetry breaking" provided the solution found at the end of the 1960s.

3.4.1 Spontaneous Symmetry Breaking

Several theoretical physicists have developed about 40 years ago a scheme which allowed for a mass of the messenger particles without destroying the symmetries of the interaction. The mass of the particles was not put explicitly as a fundamental quantity into the equations of motion, but should emerge "spontaneously" from certain processes. To this end an additional elementary field, the so-called Higgs field was introduced (named after Peter Higgs, who was the first to consider a simple model of this type). In the state of lowest energy (the true vacuum state) the Higgs field should have a nonzero value. This condition destroys a fundamental symmetry of the theory, because in a completely symmetric vacuum state the value of the Higgs field would be zero. A scheme is set up such that the symmetric state the "false" vacuum is no longer the state of lowest energy. In the theory then the masses of the particles are all proportional to the value of the Higgs field in the true vacuum state. The dynamics itself, i.e., the equations of motion remain fully symmetric, only the new state of lowest energy becomes nonsymmetric. Thus the so-called spontaneous symmetry breaking has the effect that the solutions no longer have all the basic symmetries of the equations. This aspect can be illustrated nicely with analogies from other fields of physics. In Fig. 2.19 we have already presented an example from classical mechanics. Look again at the spherical ball on top of the hat-shaped surface, as drawn in the figure. This state of the system is obviously symmetric against rotations around the vertical axis through the ball and the top of the hat. Gravity is supposed to act only in the vertical direction too, and obeys therefore the same symmetry. In this configuration the state of the system (the position of the sphere) shows the same symmetries as the fundamental forces. This state, however, is not stable. A small shift in the position of the sphere makes it roll down, and by frictional effects come to rest somewhere in the brim of the hat. A stable new configuration has been reached, but the rotational symmetry is no longer there. Although the actual location of the ball on the hat's brim is not important, any such state will break the original symmetry. The gravitational force still has the original rotational symmetry, but not the actual configuration – a situation which has been termed "spontaneous symmetry breaking."

   In the case of field theory one imagines a similarly shaped hat-like surface (Fig. 3.5) denoting the states of different energy of the Higgs field. (ϕ stands for the value of the Higgs field in a certain state.) The state in which $\phi = 0$ corresponds to an invariant

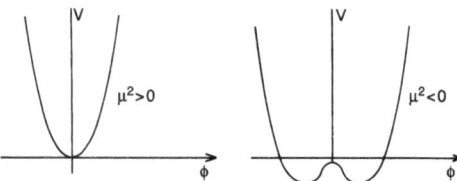

Fig. 3.5 The schematically drawn potential V of the scalar field ϕ has for high temperatures or energies ($\mu^2 > 0$) a stable minimum at $\phi = 0$, as can be seen in the *left-hand panel*. With lower temperatures ($\mu^2 < 0$) two new minima appear at nonzero values of ϕ. The state $\phi = 0$ becomes metastable, and a small perturbation pushes ϕ from this "false" vacuum into the stable "true" vacuum. This breaks the symmetry of the theory which holds for $\phi = 0$ (*right-hand panel*). The value of ϕ in the stable vacuum gives rise to masses for the previously massless particles. These masses are proportional to the nonzero value of ϕ in the true vacuum

configuration of the system. The shape of the energy surface $V(\phi)$ changes with the energy regime, where the interactions are happening. At very high energies the symmetric state $\phi = 0$ would be stable (Fig. 3.5, left panel), while at lower energies the new stable ground state, where ϕ is not zero, becomes accessible.

In the early universe the Higgs field could initiate a "phase transition," and a spontaneous symmetry breaking, when the cooling by the expansion of the universe would lead to a change of the energy surface as in Fig. 3.5. This leads to the interesting conclusion for the early universe that initially all the elementary particles are massless. Only by expansion, subsequent cooling, and spontaneous symmetry breaking by a Higgs field, the masses are generated.

3.4.2 The Electroweak Interaction

A theory of weak interactions can be formulated in analogy to the electromagnetic theory. One essential difference is the richer system of symmetries. This requires the introduction of three messenger particles, named W^+, W^-, and Z^0 boson. $W^\pm$ carry the electric unit of charge $\pm e$, while Z^0 is neutral. Combining the weak and electromagnetic interactions in one scheme, the photon is added as a fourth messenger particle. By the Higgs mechanism for this combined "electroweak" theory the W and Z^0 bosons

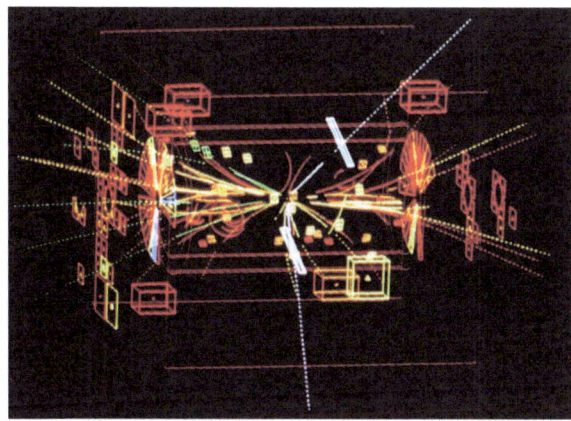

Fig. 3.6 The first picture of an event involving the neutral Z^0 particle of the electroweak interaction. The experiment was carried out on April 30, 1983 at CERN in Geneva. Among many other tracers of particles one can see in the *lower, right-hand half* of the picture sharply converging yellow lines. These are the curved trajectories of the electron and the positron generated by the decay of the Z^0 particle (*courtesy of CDS CERN*)

acquire a mass, while the photon stays massless. The introduction of a suitable Higgs field led to the prediction for the mass m_W of the W boson:

$$m_W = 83 \pm 2.9 GeV.$$

A mass for the Z^0 was also predicted:

$$m_Z = 93.8 \pm 2.9 GeV.$$

It was a spectacular success of the electroweak theory, when these predictions were confirmed in proton–antiproton collision experiments at CERN near Geneva in 1983. The messenger particles of the electroweak theory were discovered in these experiments (Fig. 3.6) with masses in perfect agreement with theoretical predictions:

$$m_W = 80.6 \pm 0.4 GeV.$$
$$m_Z = 91.6 \pm 0.03 GeV.$$

This result which brought the Nobel prize for C. Rubbia and J. van der Mer in 1984 naturally increased the confidence in the theoretical idea to describe the interactions of the elementary particles by gauge theories plus spontaneous symmetry breaking.

What is still missing is the discovery of the Higgs particle. The hope is that physicists will find it using the new accelerator at CERN, the Large Hadron Collider (LHC) which went into operation in 2009. The Higgs particle will be difficult to detect, because its mass cannot be predicted. It is, however, a most fundamental ingredient of the electroweak field theory.

The Higgs field shows one way how to introduce particle masses, and it defines a new ground state. In addition, it acts even at the smallest distances like an all-pervading, basic medium. It somewhat reminds one of the ether, the mysterious medium postulated in the nineteenth century to explain the propagation of electromagnetic waves.

Another prediction of the electroweak theory concerns its behavior at small distances or very high energies: When the energies of the reactions increase, the differences between the electromagnetic and the weak force become smaller, and at energies large compared to the mass of the W and Z bosons, the two forces merge into a single one.

3.4.3 The Strong Interaction

Another promising step on the way to a unified theory of elementary particles was the construction of a theory for the strong interaction along the lines which had led to the successful quantum theory of electrodynamics. The theory was named quantum chromodynamics (QCD). It explains the force which binds protons and neutrons in the atomic nucleus as resulting from the forces between the quark constituents. Proton and neutron are not elementary particles, according to the standard model, but complex bound states of quarks. Forces between the quarks are transmitted by messenger of mass zero, the gluons. Gluons have the peculiar property that the force they transmit becomes extremely weak at small distances, the quarks in the nucleon then behave like free particles, they become "asymptotically free." At distances of the size of the nucleon the interaction becomes stronger with increasing distance, and keeps the individual quarks and gluons inside the nucleon. In contrast to the property of asymptotic freedom this "confinement" of the quarks could be formulated only in simplified models. This formulation brings the great

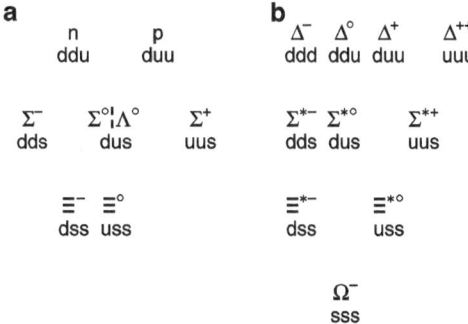

Fig. 3.7 The heavy elementary particles, the "baryons," can be arranged in an octet (**a**) and a decuplet (**b**)

reward that well-defined approximations become possible which allow us to compute in detail various reactions. The proposition of a substructure of quarks for the hadrons has turned out to be a very helpful means to understand the processes of strong interactions.

In Fig. 3.7 the octet and decuplet representations of the baryons with the various known hadrons and the way they are composed of quarks is shown. The electric charge of quarks must be 1/3 or 2/3 of the elementary unit e, since the sum of 3 quark charges must for instance be 0 for the neutron, and 1 for the proton.

3.5 The "Standard Model"

Altogether these unified theories of the electroweak and strong interactions lead to a satisfactory description of the known facts of elementary particle physics. Experiments have not yet indicated that corrections to this model are needed, except for the evidence for nonzero neutrino masses which requires a slight extension, but not any fundamental change. The community of particle physicists hopes that this will change, when the first results from the LHC are obtained.

Measurements of decay reactions of the Z^0 boson brought the important insight that there are no more than three types of neutrinos. These must be the known ones, the electron-neutrino, the μ-neutrino, and the τ-neutrino. All the six fundamental

Table 3.4 The fundamental particles of the standard model

Three families of leptons and quarks				Charge
Leptons	$\binom{\nu_e}{e}$	$\binom{\nu_\mu}{\mu}$	$\binom{\nu_\tau}{\tau}$	0 −1
Quarks	$\binom{u(up)}{d(down)}$	$\binom{c(charm)}{s(strange)}$	$\binom{t(top)}{b(bottom)}$	1/3 −2/3

Messenger particles: photon, $W^{\pm}$, Z^0 (electroweak interaction)
Eight gluons (not directly observable; strong interaction)
Higgs Particle: electrically neutral

species of quarks can also be arranged within this structure of three families. Then for the fermions and bosons in the standard model there is an arrangement as in Table 3.4.

In addition it has been useful to divide the fermions up into left- and right-handed states. Right- and left-handedness refers to the orientation of the spin with respect to the momentum. This so-called helicity is positive, if spin and momentum point in the same direction. It is furthermore a conserved quantity. This leads to the somewhat strange looking construction that, e.g., the left- and right-handed components of the electron are counted as two different fundamental constituents. In total a fermion family of the standard model consists of 15 elementary fermion states.

3.6 Grand Unification

QCD and electroweak theory show a guiding principle how to describe subnuclear phenomena in a unified formalism. Despite all its successes the standard model leaves some things to be desired – theoretically at least. The introduction of the Higgs field and its coupling to other elementary fields is essentially not fixed by the theory, quasi put in by hand. Many masses and other parameters of the standard model are not determined by the theory, but taken from experiment. There are also a number of problems which hint at the necessity to look for a more fundamental theory.

Quarks and leptons have many similarities, both are elementary quantities without any inner structure. Each lepton family (as ν_e; e) has relatives among the quarks (as u,d). Can we expect that, almost in analogy to the electroweak theory, the strong and electroweak interactions merge into a single one at very high energies?

It seems quite straightforward on the way to a Grand Unified Theory ("GUT") to treat quarks and leptons on an equal footing, and to embed each fermion family into a more comprehensive family of elementary objects. The simplest case would be to combine quarks and leptons from one family to a "grand family" of the new theory. This solves already the mystery why the electric charge appears only in specific fractions of the elementary unit e: It is a necessary consequence of the family reunion.

In a GUT particles can be transformed into each other. Some of these processes are already known, others, however, are new, like the transformation of leptons into quarks. These reactions are introduced by the messenger particles of the new fundamental symmetry. Such force-transmitting particles of the new GUT are commonly designated as X and Y bosons. In the simplest GUT they carry an electric charge of $\pm\frac{4}{3}e$ and $\pm\frac{1}{3}e$.

Some of these reactions induce a decay of the proton, the particle which is completely stable in the standard model. Since these processes are brought about by the hypothetical X and Y bosons, the decay rate depends sensitively on their mass. It is evident that the masses of the X and Y bosons must be very large, because in all experiments carried out so far, the proton appears to be stable.

In a GUT you can accurately compute how the effective strength of the interaction changes with energy. Such a behavior is shown schematically in Fig. 3.8:

Strong, weak, and electromagnetic interactions become comparable at the huge energy of $10^{15}\,GeV$. The initial data, the relative strength of the interactions at energies accessible in experiments, are determined from precision measurements of weak interaction processes, and from the masses of the W and Z bosons. Above this large threshold of $10^{15}\,GeV$ there is only one interaction, and its symmetries are spontaneously broken at lower energies. But, in fact, the unification seems to be even more complex. More recent measurements indicate that the different interactions do not exactly meet in one point. The real unification

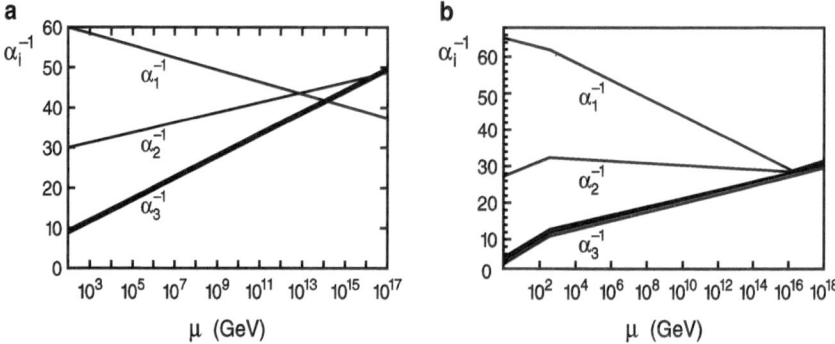

Fig. 3.8 Electromagnetic, weak, and strong interactions change with increasing energy according to GUT models such that at very high energies they converge to a single fundamental interaction. (a) Simple GUT; (b) Supersymmetric GUT

would then occur at much higher values of the energy, maybe even as high as the Planck energy $M_{Pl} = 10^{19}\ GeV$ (Fig. 3.8).

The spontaneous symmetry breaking of the GUT provides masses for the X and Y bosons of the order of the energy threshold $M_{GUT} \sim 10^{15}\ GeV$. For the hypothetical decay of the proton this means a long decay time, i.e., a lifetime of at least 10^{30} years, much longer than the age of the universe. Since a single proton would very likely not show any effect in a reasonable time span, a large number of protons were used in experiments. In huge underground water tanks in Japan and in the USA up to 8,000 tons of purified water (about 5×10^{33} protons) were watched by photo electrodes. Not a single significant decay event was found. Thus a lower limit of 10^{32} years for the lifetime of the proton has been established. This value already contradicts the predictions of the simplest GUTs. So far the proton decay experiments have been the only tests of GUTs. The energies reached in accelerators (10^3 GeV for the LHC) are frustratingly far from the GUT unification energy. But in the earliest phases of the universe thermal energies can be as high as GUT energies, and higher, if the hot big-bang model is correct. Any hypothetical GUT can at least be tested then with respect to its consistency with a hot initial state of the universe.

3.7 A Theory of Everything

3.7.1 Superstring Theory

We have mentioned already in Sect. 3.1 the theory of "super-strings" which starts from the idea that the smallest building blocks of the world are not point-like elementary particles, but even more fundamental structures which resemble string-like, vibrating concentrations of energy. These strings are 10^{20} times (in words hundred billion billion times) smaller than an atomic nucleus. Just as the string of a violin can vibrate in different modes (each one producing a different tone), the tiny string can oscillate according to superstring theory in many different ways. These vibrations ought to produce the various properties of elementary particles. The tiny string oscillating in a specific way might appear like an object with the typical mass and charge of an electron. It would then be the object that commonly is called an electron. Vibrations in other patterns would have the properties required to identify them as a quark, a neutrino, or any other elementary particle. All types of particles are unified within the superstring theory, since each one corresponds to a specific vibrational pattern of the same fundamental unit. Even space and time ought to be represented in this way eventually.

This is a very beautiful mathematical entrance door to a theory of everything, and perhaps, after all, it is the true theory. Up to now, to be sure, one has not succeeded to reconstruct the known elementary particles, let alone a realistic space–time within the context of string theory.

The great hope of string theorists is to achieve the long desired unification of quantum theory and general relativity with the help of their conceptions. There are positive signs, but the goal has not yet been reached.

The vibrating strings are described mathematically in a 10- or 11-dimensional space–time, and the way back to the common-place four-dimensional world is a complicated procedure which should guarantee that the surplus dimensions stay microscopi-cally "small" in some sense. They are "rolled up" in a microscopic structure. In the discussion of the singularities at the big bang and inside black holes, I have already made use of the illustrative picture of a straw. At great distance it appears one-dimensional like a piece of a line, a closer look reveals the two-dimensional tube, and a very close scrutiny the finite thickness of the wall of the tube, i.e., the three-dimensional structure. One imagines that in close analogy the rolled-up dimensions show up, if one looks into very small scales of space–time of the order of the Planck length. String theory predicts the existence of a highly complex six- or seven-dimensional space at each point of space–time. Since there are many such spatial structures, there are also many solutions of string theory. Apparently too many to find the one fitting to our world by trial and error. Perhaps our world is not even contained in the manifold of solutions? It will be a hard piece of work for string theorists to obtain a definite result. The reward would doubtlessly be considerable: Since in string theory only mathematically well-defined quantities occur, the divergencies of quantum field theory, and the singularities of black holes and the big bang would be absent. It would really be interesting to see the string solution describing a nonsingular big bang, or to look inside a black hole and see the concentrated energy packet which has replaced the singularity.

Until that time we will have to wait patiently for quite a while. At the moment one cannot foresee the consequences of this theory for our normal world or for the understanding of the early universe. String theory remains an intellectual enterprise which has not yet found its place in our scientific view of the world.

Clearly, superstring theory can only yield results which can be tested by experiment, if a low-energy approximation is constructed. "Low energy" means at least within the energy scale of a GUT theory, because the standard superstring theory describes the situation above the Planck energy of 10^{19} GeV. More exotic versions have been formulated which may even lead to experimental signatures at the scales reached by the LHC. It would be very exciting, of course, if any of these speculative features were actually detected.

An enjoyable book on this topic is Superstrings (edited by P. Davies, J. Brown), where extensive interviews with fans and opponents of this theory have been recorded. A presentation of the theoretical concepts in parts understandable by the nonexpert can be found in the book by Brian Greene *The Elegant Universe*.

3.7.2 Stable Fields and Volatile Matter

Let me close this rough survey of the world of elementary particles with a short comment.

We have seen that the elementary particles in the standard model can be arranged in three families. Quite remarkably, the first family is completely sufficient to build up all the chemical elements, the molecules, and everything else of our normal world. The two additional other families are not necessary for that, they form an ornate attribute of some high-energy reactions. Why is this uneconomical, additional expenditure of two other families of particles there? The physicists do not know.

Another important aspect is the fact that material particles as well as fields exist as real, physical objects. Fields are immaterial, not substance, but pure form. Nevertheless, they propagate in empty space without any material carrier, they possess energy, and they play a fundamental role in the interactions of elementary particles. In fact, fields seem to be the really fundamental structures of the world, and not the elementary particles which change into one another or into energy, which decay or are created anew from excitations of the fields.

It is a strange picture which emerges here at the very foundations of everything. Material objects are volatile and changeable, permanent existence is a property of abstract structures like fields or strings.

Our world only seems to be solid, but in truth it stands on shaky, uncertain foundations. What would happen, if the potential of the Higgs field were to change suddenly by a tiny bit? The masses of the elementary particles would also change, and perhaps in such a way that atoms could not exist any more. Then, at a glance, everything would have disappeared, our whole beautiful world, and we too.

Of course, such theoretical concepts are preliminary hypotheses. On the other hand, we must acknowledge that those descriptions and considerations are not pure fantasy, but speculations based on solid experimental findings.

The basic forms and structures, respectively the ideas which are currently fashionable, cannot be exactly presented in everyday language, but only in pictures and analogies which are never completely accurate. We need the help of a mathematical formulation to describe these notions consistently. String theory which is claimed to be the fundamental theory for everything is especially attractive to many researchers, because of its mathematical structure. Experimental tests of the theory do not appear to be very promising, and the prospects of success of the whole scheme are uncertain. A bet along the lines of Pauli should not be risked. We will not bet that the all-comprehensive theory will never be found. We could never win such a bet, even if we were right. Whether a successful theory can be developed without any guidance by experiments remains to be seen.

After this big journey in search of what the world is made of, we will address the question how quantum world and classical world are connected.

3.8 The Strange Reality of the Quantum World

3.8.1 Particle, Wave, Field

At the level of the elementary particles the last traces of individuality are lost which are still present at higher levels of complexity. Any arbitrary electron or proton is exactly the same as any other. Therefore it is correct to talk of "the" electron, "the" proton.

Elementary particles are characterized by certain unchanging properties like mass and electric charge. They behave like small bodies obeying in their reactions the conservation laws for energy, momentum, and other specific quantum numbers, when they collide at high energy in the experiments at the large accelerators. The analysis of scattering experiments at high energy is completely and satisfactorily carried out in this picture of "corpuscles."

But as quantum mechanical objects the elementary particles also have properties which contradict the image of small bodies drastically. They possess in addition the typical properties of waves which have been demonstrated for example in many interferometric experiments with electrons and neutrons. A particle of mass m has a characteristic wavelength, its "the Broglie wavelength"

$$\lambda = \frac{h}{mv},$$

and this correspondence is not just formal. (h: Planck constant; v: velocity; m: mass). When they are scattered off a grid with a distance less than λ between the slits, the particles behave like waves with wavelength λ. On the other hand, it has turned out, starting with the work of Max Planck around 1900, that the electromagnetic radiation itself, such as the light waves, is made up of discrete quanta of energy which were called photons. Photons have spin 1 and mass zero. An object of mass zero may not seem very tangible. But these particles of light demonstrate their particle nature in experiments such as the photokinetic effect, where the irradiation of a metal plate by light produces an electric current. This is achieved by the photons which kick electrons out of their atomic shell. The so-called dualism of wave and particle is one of the fundamental facts of quantum mechanics, although it seems to be self-contradictory and paradoxical. It indicates that neither the classical notion of the particle nor the concept of a wave fully covers the reality of these quantum mechanical objects. We shall discuss this aspect further in what follows.

3.8.2 Deflection at a Slit

The fundamental experiment, where the wave character of elementary particles has been demonstrated, is the double-slit experiment of Young. The principal issue is to register the electrons of a beam which have passed through two narrow slits in a metal screen. With both slits open a pattern of stripes appears on the detector screen corresponding to the interference pattern of light waves passing through a grid, while in fact the stripes just show the distribution of electrons hitting the detector. With one slit closed the electron distribution at the detector fits a bell-shaped Gaussian curve (as it had been pictured on the 10-DM bills in Germany) which results from the random scatterings and subsequent deflections by small angles of the electrons passing through the slit. The same beam of electrons thus behaves sometimes like a wave, sometimes like a stream of particles. It is the arrangement of the experiment which decides on the outcome (see Figs. 3.9 and 3.10).

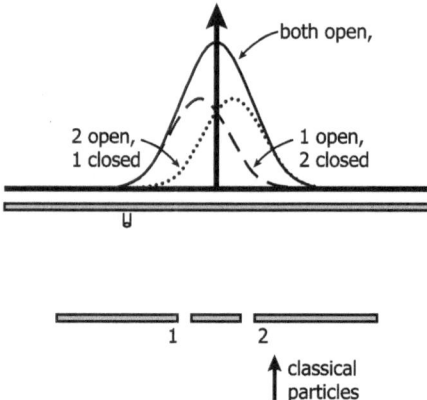

Fig. 3.9 The passage of classical particles through two small slits in an otherwise impenetrable screen produces hits at a detector surface behind, which are distributed in the shape of a Gaussian curve, exactly as it is expected, if small random changes of the directions of the particles occur. If one slit is closed, the Gaussian curve for the open slit is produced

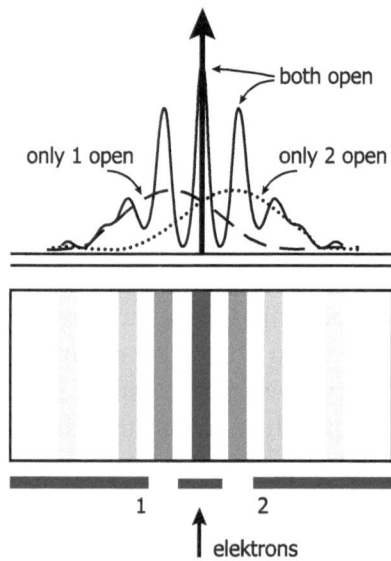

Fig. 3.10 The passage of quantum mechanical particles (electrons, photons, and others) leads to an interference pattern of the electron hits (indicated by *black and white stripes*) at the detector, if both slits are open. If only one slit is open the Gaussian curve shows the distribution of particle hits

3.8.3 The Change of the Particle Concept: Heisenberg's Uncertainty Relation

In classical physics the momentaneous state of motion of a particle is described by two independent data, namely its location and its velocity. In quantum mechanics the situation is somewhat different: Location or velocity can each in itself be determined with arbitrarily high precision, if one does not care about the other quantity. Both quantities, however, cannot be determined simultaneously with arbitrary precision. It is not even permitted to imagine that at the same instant they both have well-determined values. Position and velocity are actually uncertain to some degree. The German physicist and Nobel prize winner Werner Heisenberg has formulated an "uncertainty principle" which is one of the foundations of quantum mechanics: There is an uncertainty in the position of a particle Δx, and an uncertainty in the momentum Δp (momentum is just velocity times mass). If we multiply Δx by Δp, the product is never smaller than $\hbar$, Planck's constant, a constant of nature which is present in all quantum processes. Take for example an electron, and measure length in centimeters, and time in seconds. Then the numbers are such that a variation of the velocity in a range of $1\,\mathrm{cm\,s^{-1}}$ implies an uncertainty in the position by 1 cm. If you want to know the position more precisely, the velocity becomes more uncertain, and vice versa. That is strange, isn't it? But Heisenberg's uncertainty principle is a fundamental insight, a basic property of the quantum world. Why is it so remarkably different from our common experience? We shall emphasize that this just cannot be explained further, for instance by falling back on classical concepts, but that it must be accepted as a property of nature.

3.9 What Is the Reality Described by Quantum Theory?

Elementary experiments lead, as we have just seen, to the conclusion that the quantum world is different from the classical world. In the classical world we set up our detectors, here we register the results. But an interpretation in classical concepts leads to the strange result that the quantum objects sometimes behave like particles, sometimes like waves, depending on the experimental setup. The results of the experiments are in general not fixed as in a deterministic process, but there are usually several possibilities for the particles, respectively waves to behave. In each single test case they decide to choose one of these possibilities randomly, and unpredictably.

A large number of quantum processes then results in a distribution, where each possible result appears with a certain probability. The quantum mechanical laws seem to allow some degree of freedom or possibilities of choice.

Thus a given atomic nucleus of a radioactive element, let us say uranium, will decay eventually, but for the single nucleus it is impossible to predict when this will happen: It can happen immediately or in a billion years. A sparrow on the roof also has a finite lifetime, and his life expectance can be estimated much more exactly than that of the uranium atom.

Each classical object consists of very many atoms. Many quantum objects packed together obviously lose their quantum properties, and the total assembly behaves classically. How this transition to the classical regime is happening, and where the boundary between quantum and classical world lies, has yet to be clarified.

Erwin Schrödinger, Austrian cofounder of quantum mechanics, has illustrated the curious situation by his famous example of the cat in a box, into which a poisonous gas can flow. (Cat lovers please excuse this terrible picture! Schrödinger himself was actually very fond of cats.) The gas would kill the cat. The gas flow is released by the decay of a radioactive atom – by a quantum mechanical process. The complete system – cat, box, gas, radioactive atom – can be described as one quantum mechanical system. This system can have many different states which are, however, not immediately realized, but are all kept in a kind of superposition. The quantum mechanical "wave function" containing all those states in linear superposition is the full description of the system. This function obeys the so-called Schrödinger equation which determines the time evolution of the wave function. Among the states there is also a linear superposition of the alternatives A: *Cat is alive* or B: *Cat is dead*. If you let the experiment run for a while, and then look into the box, either A or B has occurred. But, if nobody checks, and the system is left to itself, everything remains in a linear superposition of all the states. The cat, in a sense, hovers in a strange interim state between A and B, between dead and alive.

This sounds completely absurd, but such is quantum mechanics, if we try to relate it to the classical world.

Let us look again in some detail at the experiment, where an electron beam passes through one or two slits in a metal screen, and then impinges on a surface, where the electron hits are registered. If only one slit is open, one finds an intensity distribution at the detector surface which resembles the bell-shaped Gaussian curve indicating randomly occurring deflections of the electrons, when they pass the slit. If both slits are open, the intensity distribution at the detector does not at all resemble the sum of the intensities of two single slits, but a pattern of stripes is formed, an interference pattern. Remarkably, the number of electrons which hit a unit surface area of the detector is much larger in the stripes than in the one-slit experiment. On the other hand, there are locations between the stripes, where no electron is registered, although in the case of one open slit, there were electron hits at this location. For the electrons there obviously are paths which cannot be travelled, when two slits are open. A strange result, since according to our naive expectation, there should be more possible paths from the source to the detector screen, when two slits are open. The result evidently is due to the interference properties of waves. When two slits are open, the electron beam behaves wave-like, when one slit is open it behaves like a stream of particles (Figs. 3.9 and 3.10).

The intensity of the electron beam can be reduced until it is certain that only a single electron is travelling between source and detector at any one time. Even then the same outcome is found: bell-shaped Gaussian, when one slit is open, interference pattern, when two slits are open. This must mean that the single electron can act either as a wave or as a particle. It carries both properties which appear to us mutually exclusive. If we place a detector at the slit to register whether an electron has passed or not, the interference pattern disappears, and the Gaussian random distribution of the superposition of the individual electron hits turns up. We can arrange the experiment such that the two slits are separated so far from each other that the particle concentrated in a small volume and passing through one of the slits definitely cannot take notice of the other slit, whether it is open or closed. But the experiments always demonstrate that the particle knows about the situation, and behaves accordingly. How can that be? Obviously we must assign to the electron a kind of spread out existence, covering the whole spatial region – including metal screen, detector, and source.

The whole story acquires a further strange twist, if the experiment is set up such that one can open or close the slits, after it is absolutely certain that the electron has already passed through. If at first only one slit is open, the Gaussian distribution should result. But if the second slit is opened, *after* the electron has passed the metal screen, the interference pattern shows up. Clearly, no thing in our classical world could achieve such a trick! Electrons are not of this world, it seems.

Anton Zeilinger and his colleagues in Vienna have set up sophisticated experiments which allowed them to try the double-slit experiment on larger and larger molecules. Meanwhile the interference pattern has been demonstrated for molecules consisting of more than ten million atoms. The classical behavior apparently shows itself only with still larger objects. But why does it show up at all?

3.9.1 A Few Mathematical Remarks

It is the place now to speak about a few mathematical aspects of quantum mechanics. The following paragraph is no more than a very basic introduction to the mathematical concepts behind the illustrative pictures described so far. One way to a mathematical description of quantum objects is to construct solutions of a differential equation which has been formulated by Schrödinger, and which is rightfully called "Schrödinger equation." The "wave functions" which are solutions of the Schrödinger equation correspond to certain quantum states. This might be, for example, an electron having its spin oriented in a definite direction, or an electron with the opposite spin. The combination of two electrons with spins in opposite directions would then also be a solution, another possible quantum state.

If there are two solutions ϕ_1 and ϕ_2, then all linear combinations $a\phi_1 + b\phi_2$ with complex numbers a and b are also solutions. The complete wave function contains all these possible states. Even the coexistence of alternatives is not excluded, i.e., the states which could not exist simultaneously in the classical world. Thus the state, where the electron passes through slit 1 in the double-slit experiment, and the state, where it passes through slit 2 (cf. Figs. 3.9 and 3.10), may be combined to a curious state, where everything remains strangely undecided in-between. Clearly, the Schrödinger wave function cannot be interpreted as a real wave, or as corresponding to the picture of the "matter wave" discussed before. How then does the interference pattern arise? The situation is as follows: In quantum theory the wave function evolves completely deterministically as the solution of the Schrödinger equation, a differential equation. But it does not describe any real wave or state. A definite state is realized with a certain probability, and this probability is given by the absolute square of the complex numbers. These numbers are called probability amplitudes. Take, for example, the state consisting of a superposition of the alternatives for the electron to pass through slit 1 or slit 2. The measurement of the electron passage through one of the slits reduces the superposition of the alternative states, and "lifts" one alternative into reality. The measurement of many electrons then tells us that the hits at the detector are distributed on the surface according to a probability which corresponds to the absolute square of the probability amplitude of this alternative. When we set up the experiment such that not the passage through the slits, but only the hits on the detector screen are registered, then in the superposition of alternatives the interference of "in-between" states leads to a more complex probability distribution, and the result is the interference pattern of the double-slit experiment.

The prescriptions of how to compute the result of a measurement work extremely well for practical purposes. In fact, quantum mechanics allows us to compute subtle effects very precisely, and in perfect agreement with experiments.

Difficulties arise, when one tries to interpret these prescriptions.

3.9.2 Attempts to Interpret Quantum Mechanics

In principle one can hold one of two possible opinions: One could be content with the fact that classical and quantum objects behave quite differently. The interpretation of a measurement, that is of a classical event – like the direction of a pointing needle, the luminescence of a tube filled with gas, or the electric current in a detector – must then be done in pictures which apparently contain contradictions, but only because our common, classical language is not adequate for quantum mechanics. This interpretation of quantum mechanics has been developed by Niels Bohr and his colleagues in Copenhagen in the 1920s and 1930s. It is still accepted as a reasonable way to understand the meaning of quantum mechanics by most physicists.

According to the so-called Copenhagen interpretation the measurement is itself an essential element in the formulation of the theory. The measuring process is a serious intervention. The undisturbed wave function evolves according to the Schrödinger equation in a completely deterministic way starting from definite initial conditions. But the measurement and the subsequent "collapse of the wave function" introduce a nondeterministic element: The sudden transition to a result which can be recognized in the classical world happens, because an observer looks at the outcome of the measurement. As long as nobody looks everything remains in the quantum mechanical in-between state.

Werner Heisenberg, who was a coauthor of the Copenhagen interpretation commented as follows:

> It has turned out that this objective reality of the elementary particles which we had hoped for, is too coarse a simplification of the true situation, and must be replaced by much more abstract notions. If we want to obtain a picture of the elementary particles – what they really are, and how they behave – we can no longer disregard the physical processes implied in our attempts to learn about them. Each observation causes a gross disturbance of the behavior of the smallest building blocks of matter. It is impossible to speak about

the behavior of the particle separated from the measuring process. This has the consequence that the laws of nature which we formulate in quantum mechanics do not apply to the elementary particles themselves, but to our knowledge of the elementary particles... The idea of an objective reality of the elementary particles has thus evaporized in a strange way, not into the fog of some new, unclear or not yet understood concept of reality, but into the transparent clarity of the mathematical formulation which ... represents our knowledge of the elementary particles.

The poet Christian Morgenstern has expressed a similar view in his verse about the mile post: "...what is it, not seen by us, unknown, non-existing thus. Just the eye creates the world." ("...was wohl ist er ungesehen, ein uns völlig fremd Geschehen/ Erst das Auge schafft die Welt.")

Many scientists accept the Copenhagen interpretation as useful for practical purposes, but they also see that here a new concept is introduced into the theory, not clearly defined and somewhat vague. The fact that a conscious observer is included in the experimental arrangement means a dramatic change.

Quite in contrast to the original approach in science to describe the world which exists objectively and independently of the observer, one would have to consider reality in a certain dependence on the mind of a conscious observer. Has the attempt to achieve an objective picture of the world been wrecked already at this level of quantum mechanical processes?

A main difficulty with the Copenhagen interpretation lies in the separation of the observer and the system. The more one analyzes this in detail, the more the issue becomes ambiguous. Look at the observer, for instance, reading off the position of a pointer – his glasses can be included offhand in the system, as well as the physical–chemical reactions in his optic nerve up to the final information-processing in the brain. Up to that point everything can be added to the quantum mechanical system to be measured. The state hovering in the superposition of various alternatives would remain. Only the act of consciousness, when the conscious observer perceives the result, leads to the collapse of the wave function, to the real, classical event. It seems, as if this interpretation opens a window to consciousness, to the subjective mind, an area which should remain closed to physics by definition according to the scientific method.

Quite a number of physicists attempt to bring into physics at this point concepts like mind, consciousness, free will, even God. I would be absolutely enthusiastic, if there were really a possibility to demonstrate that the world view of physics, starting from an explanation of electrons and atoms would eventually end up at such "nonobjective" conceptions as mind and consciousness. But the whole argument is by no means compelling: The interpretation of quantum mechanics in terms of "measuring process," and "observer" appears more like a makeshift which indicates that the theory is perhaps not yet complete, and that it must be replaced at least in this aspect by a more comprehensive description. This does not exist yet, but that does not mean that it will not be found within a few years. All the playful speculations which aim at deriving ingredients for metaphysical or theological arguments from quantum mechanics will be spoilt by the new theory. It is my opinion that science can honestly only lead us up to a threshold at which it becomes evident that further explanations or arguments must transcend science. What science can find and achieve is interesting in its own right, as we have seen. The complexity of the world becomes more and more fascinating, as we discover more and more about it. But there are many things of great importance which are left aside by the scientific method.

On the other hand it seems that quantum mechanics really describes atoms, electrons, quarks, and strings, and not primarily the special macroscopic events connected with the measurements of what we call "the properties" of these things. But if these objects cannot be identified with the wave function in some way and if to speak about them is not just a shortcut for complicated statements about measuring processes, then one must ask where one can find them in the quantum mechanical description. Maybe there is a very simple reason, why it is so difficult to recognize in the mathematical framework of quantum mechanics the objects which ought to be dealt with? Maybe the quantum mechanical description is not the whole story?

It was Albert Einstein, who believed that despite the great successes in the computation of many subtle effects quantum mechanics must be an incomplete description of reality. All the attempts to find an objective representation of quantum mechanics are, however, up to now quite unsatisfactory and somewhat strange.

Among these attempts the many-world hypothesis is extremely bizarre: This interpretation assumes that each quantum mechanical effect splits the world into a manifold of parallel worlds, each one the home for one possible result of a measurement. The wave function is viewed as a real quantity which exists, however, in different worlds. Such an explanation buys a realistic interpretation of the wave function at great expense: At each moment incredibly many such splits occur, and new universes arise in gigantic number. There is no communication at all between these worlds, and we ourselves do not feel that we are permanently duplicating, because our conscious self apparently glides totally unaffected on a single path through these worlds which split up into new ones incessantly.

In the many-world hypothesis the splitting-up of the worlds happens in an absolutely deterministic way, guided by a super-wavefunction which never collapses, except if a God outside of the universe would observe it.

Many feel more sympathy for the idea that there is a kind of "underworld" of quantum mechanics, a hidden reality which we cannot observe. There, deterministic laws rule, and statistical fluctuations lead to the quantum mechanical phenomena which we observe. We shall see in the following that such a view of quantum mechanics can be tested experimentally. This has been done, and the idea of a hidden, deterministic world has been disproved.

3.9.3 The EPR Paradox

Einstein has designed a "thought experiment" which considers the case where a particle and its antiparticle are created simultaneously, and then propagate away from each other in opposite directions. The pair of particles has one quantum mechanical wave function, and therefore they remain one system, they stay "correlated," even if they are separated by large distances. Measurements carried out for one of the particles cause simultaneously a measuring result for the other, even if they are light-years away from each other. Let us look at the spin of the particles, and its various possible orientations in space. All these spin directions are contained in the wave function, and each one can be measured with a certain probability. The wave function is reduced by a measurement to a state with a definite spin orientation, it "collapses." Both the particle and the antiparticle do not have a definite spin orientation initially, but they are in a state where all possible orientations are contained in linear superposition. But with the measurement of the spin of the particle, the antiparticle simultaneously obtains a specific value for the spin, namely exactly opposite to that of the particle. This happens even if particle and antiparticle are so far away from each other that even with light signals there is no possibility for them to communicate. Since neither the particle nor the antiparticle has a definite spin orientation before it is measured for one of them, the question arises, how the other knows which spin direction it should have.

This argument was sharpened by Einstein, Podolsky, and Rosen in 1935 in the so-called EPR paper to a paradox. The "EPR Paradox" consists in the fact that despite the causal separation of the measurements at the two particles there is obviously a "spookish" noncausal communication which ensures the adjustment of the spins. Therefore, Einstein argued, there must be a classical world, unobservable for us, which causes the quantum phenomena by statistical fluctuations of its variables.

This idea follows the usual interpretation of statistical processes in classical physics. When we toss a coin for instance, we say the result (head or tail) is random. But if we took into account the motion of the air, the exact initial conditions of the throw, and the properties of the coin, we could simply compute the result. Since we do not have such detailed knowledge, the tossing of a coin appears to be a random event. Quite similarly the randomness of quantum processes might be due to our lack of knowledge of the underworld of hidden variables. Can this theory be critically tested by experiment? Yes, it can.

3.9.4 Bell's Inequality

The late physicist John Bell, who worked at CERN in Geneva, had found in 1964 an exact estimate for certain functions, so-called correlations which can be measured for a system of two particles, and which determine the degree to which the two particles influence each other.

Classical correlations appear almost trivial. Take, for example, two things which belong to each other, like a right and a left glove. We can immediately conclude from the fact that far away from home we find a right glove in the pocket of our coat that the left one must lie at home in the drawer, unless we have lost it underway.

In quantum physics the issue is somewhat more complicated. Since "quantum gloves" do not exist, we take a pair of photons and consider its quantum mechanical correlations. None of the two photons is in a well-defined state initially. Only when a measurement is carried out on one of the photons, this acquires spontaneously one of the possible states, and at the same moment the other photon is in a definite state. If we look at linearly polarized photons, the two states could be two polarizations perpendicular to each other. We see the difference to the classical example: The right glove was always in the pocket of the coat, and the left one at home.

Bell's inequality sets a maximal value for these correlations, if hidden classical variables are the underlying theoretical concept. Bell investigated a system of two particles which are viewed as really existing, and which can only act on each other in a causal way, i.e., signals cannot propagate faster than light. These two assumptions establish what is generally called "local reality." Commonplace quantum mechanics predicts that the limit derived by Bell will be surpassed for the case of so-called entangled states, i.e., for specific correlations of the photon pair. In other words, the conventional interpretation of quantum mechanics requires a high degree of cooperation (or "spookish" conspiracy as Einstein called it) between separated particles, a property which is absent, when local reality is valid.

3.9.5 The Aspect Experiment

The lucid analysis of John Bell and various technical advances have finally led to a real experiment testing the EPR Paradox.

In 1982 Alain Aspect and his collaborators observed the emission of light quanta by calcium atoms. In the experimental arrangement pairs of photons were observed (photon A and B). The two photons in each pair propagated in opposite directions, and they were measured, when they were far apart. Since the total angular momentum of the emitted light was zero, both photons were circularly polarized. One photon has the electric field rotate clockwise, the other one counterclockwise. The observation of a certain amount of polarization at A (measured by a specific setting of the polarization filter) determined the result of the measurement at B. The outcome of the experiment showed a significant violation of Bell's inequality. The limit had been surpassed. Thus it had been demonstrated that quantum mechanics really possesses a noncausal element inexplicable in a classical sense. Even at great distances the two photons in the experiment remain correlated, they form an "entangled state."

3.9.6 Entangled States

Entangled states have correlations which surpass the limit derived by John Bell, and thus demonstrate the nonclassical, nondeterministic character of quantum mechanics. Without resorting to a mathematical formulation it is really quite difficult to describe the notion of "entangled states." For two linearly polarized photons it is a special superposition of the two orthogonal states of polarization. Perhaps the following nice brain teaser can serve as an illustration (S. Popescu, quoted by Dagmar Bruß in "Quanteninformation"): Two friends, Alice and Bob, are in the power of a tyrant, who promises to release them from prison, if they can solve in more than 75% of the cases the following problem: Two messengers shall bring both to Alice and Bob a flower. The flower has the color red or blue. Alice and Bob, who are in separate cells, must tell the messengers a different number each, e.g., Alice 0 and Bob 1, if they both receive a red flower. In all other cases their answers should agree (that is if two blue flowers, or a red and a blue flower are brought). Alice and Bob can make an arrangement, before they are separated. The best chance seems to be that Alice says "0," when she gets a red flower and "1" for a blue one. Then

Bob could say "1" for red and "0" for blue flowers. Then they could obtain the right solution in 75% of the cases on average.

But if Alice and Bob would be able to arrange for a set of entangled states, before they are separated, they could do even better. One part of the entangled state would remain with Alice, the other part with Bob. The color of the flower ought to correspond to a specific measuring prescription. Alice and Bob have determined their measurement procedure for the cases of a red or a blue flower. By a clever arrangement the correlation can be optimized, and a value of about 85% for a correct guess can be reached. Quantum mechanical correlations are larger than classical ones! Alice and Bob gain their freedom, because they know how to apply quantum mechanics in practice.

Meanwhile entangled states have been investigated in many experiments. Ideas like the quantum computer, quantum cryptography, or the teleportation of a quantum state have their origin in that concept.

Einstein's belief in a hidden classical world behind the quantum phenomena has been disproved. It would have been a way to go on viewing the world as an objective reality, which existed independently of an observer but it is no longer viable.

3.10 Transcendence?

The key role played by the observer in quantum mechanics inevitably leads to questions of the nature of mind and consciousness, and their relation to the objective things of physics.

The collapse of the wave function seems to call for the act of a conscious subject, so to speak a direct interaction of mind and matter. Since Heisenberg's uncertainty relation permits a variety of possible developments for a physical state, it seems quite seductive to postulate at this interface an involvement of mind or consciousness in the choice of the state. Thus we would need an understanding of mind or consciousness, before we could hope to make sense out of quantum mechanics.

It seems, however, that the idea that mind enters the world via the quantum mechanical uncertainty principle is not really taken seriously. The electric-chemical activity of the brain is quite robust and classical, not afflicted by quantum jumps. Neverthe-less, we cannot evade the dilemma that the concept of a real world existing independently of the observer runs into difficulties at the quantum level.

4. Boundaries and Transgressions

It is a difficult task to connect the scientific explanations of the world and our subjective experiences, wishes, beliefs, and feelings. In this section I want to explore, how a unified view of the world might be imagined, but not at the expense of declaring our feelings as illusions, and the world of physics as everything there is. If you hold on to this belief, I want to demonstrate that it is really a belief, not a scientifically proved fact. I also hope to sprinkle some element of doubt into the minds of those, who claim to believe only what they see. In fact, it is not so easy to say what it is that you see, if you delve deeply into the findings of modern science.

There are no simple answers to these age-old and still relevant questions, and I am also not able to present definite answers here. Rather I want to argue in favor of being suspicious to definite answers, and to search for the right questions to ask.

At the end we will find that the simple commonsense view of the world is shaken up somewhat, but that our beliefs are still a matter of personal decisions. Whether that can be considered as progress and as helpful is again a matter of personal belief.

4.1 Impact and Meaning of the Scientific View of the World

In the preceding two chapters we have made ourselves familiar with some basic facts of the physical world, from the largest structures in the cosmos to the smallest building blocks, the elementary particles. Already a simple inventory exhibits an astonishing variety of shapes and structures. I am always amazed at the beautiful pictures of faraway galaxies, as they are brought to us by modern telescopes. Even more impressive appear to me

G. Börner, *The Wondrous Universe*, Astronomers' Universe,
DOI 10.1007/978-3-642-20104-2_4,
© Springer-Verlag Berlin Heidelberg 2011

the findings of modern cosmology on the origin and evolution of these celestial objects: We can understand how this complex cosmos originated from an extraordinary simple initial state, how after the big bang the matter developed into a rich variety of forms and patterns. Although the physical explanations do not immediately capture the charm of the direct eye impressions, but rather reduce the colorful variety of objects to simpler and more abstract properties, I find that this does not diminish the wonder of this world. The deep inner coherence of the cosmos is almost unbelievable.

Isn't it a fantastic idea that all the stars in a galaxy with a diameter of 100,000 light-years occupy a region which close to the big bang has been no bigger than an atom? Even more remarkable sounds the history of the chemical elements: During the first minutes of the cosmic evolution the light elements hydrogen, deuterium, helium, and lithium formed. They were the raw material for the first stars which brewed the heavy elements in their interior. Every carbon or oxygen atom on the Earth stems from the interior of a massive star. When the star exploded, these atoms were thrown into interstellar space, used in the formation of new stars, and ended finally in the solar system after they had passed through several generations of stars. Thus each atom in our body has shared this history of 10 billion years. We are tightly bound into this cosmic cycle. Our material existence reaches back almost to the big bang, and we are literally made out of the dust of stars.

Furthermore we have learned that according to fundamental theories the seemingly solid real world is built upon a foundation of not so solid objects like fields and strings which are just concentrations of energy-building blocks resembling mathematical structures and ideas rather than material objects. It is worthwhile to think about this change from a simply materialistic attitude. Does it not look, as if matter was depending on mind for its existence?

We are joined to the cosmos not only by the matter we are made of, but also by the laws of physics and the constants of nature, whose actions, values, and harmonious interplay have made our existence a possibility. The considerations of the "anthropic principle" are concerned with fine-tunings which seem to be more or less self-evident: We are here, and therefore the conditions for our existence must be satisfied. But nevertheless some of these interconnections make you wonder, and we will discuss them in more detail in the following sections. One important conclusion must be stressed again: We are participating in a cosmic cycle driven by the laws of nature which govern our world. Just by this fact the view of the world as presented by science is of importance for us, despite the apparently meaningless cosmic play of atoms, stars, and galaxies. It is the perspective which matters, and since the evolution and state of the cosmic matter is the foundation of our existence, we are somehow deeply involved. Still, our basic question after the meaning of life is left unanswered. But is this after all a question which can be answered within the context of science? I believe that a sound answer based on scientific arguments cannot be given, although when considering the amazing cosmic interconnections my feeling grows that we are close to a deep understanding. Doubtlessly, when asking such questions, we have the silent hope that our real being points beyond our finite life imprisoned in space and time. But already the much easier question "Why is the universe the way it is?" takes us beyond physical experience into metaphysical realms.

It seems appealing to believe in a purpose, a plan oriented toward a goal, when we look at the cosmic evolution as it is guided by the finely tuned forces and the constants of nature. The creator of the world could have arranged everything such that the universe "becomes a hospitable place for life," as Freeman Dyson expresses it. The existence of God as creator of the world cannot be proved by scientific arguments, because the method of science to accept only objectively well-defined quantities excludes from the world view of physics all subjective structures, such as mind or God or me. On the other hand the existence of a Creator of everything can also not be disproved by scientific methods, because an almighty Creator might have set up the world with all its properties just as the physicists find it, and try to explore it.

Why then should we worry about the relations between science and religion? Isn't a clean separation the best solution? I think that this is not a satisfactory attitude, because the world is given only once, and it is the same world for all of us. Therefore we should ask in my opinion, whether we might not try to see beyond physical objectivism in science, and whether we might not be able to glimpse some objective truth in metaphysics.

I hasten to add that I am very much against the attempts to find room for religious belief and God in the gaps of our knowledge. These are very insecure accommodations: Some might disappear quickly in a new scientific development, others might become uncomfortably narrow.

Perhaps most interesting to some are the points, where faith and reason seem to clash. The origin of life, free will, the question of design, of ultimate aims of the cosmic evolution are aspects of this kind. We shall touch on these questions from both the sides of science and religion. But in a more modest way we might first try to find some common ground of science and religion. In some very basic sense there is in both areas the confidence that it is possible to uncover deep truths about the world. Scientists believe that they can learn more and more about the structure of the world by an interplay of theory and experiment, while in religion the truth is given by direct revelation or dogma. This is a very different attitude, but the belief that there is "something" out there to be discovered by us is common.

Some overlap, despite the tension between rational explanation of the world and religious belief, exists in the way both contain statements which are seemingly in conflict with common sense. I have mentioned many times, how cosmological knowledge transcends our everyday experience, especially in connection with aspects like the origin of space and time in the big bang, and their perishing inside black holes. This sets us free from the illusion that there is only the ordering of things in space and time, because if space and time themselves are subject to change, then there may also be an ordering of appearances not requiring the notions of space and time. We cannot imagine what it might be, nor can we be sure that something really exists outside of space and time, but just to think of such a possibility releases us to some degree from the tyranny of space and time. May not the clear insight into the limitations of our experiences open up a path for us to religious faith without getting into conflict with common sense and naive rational explanations of the world?

The remarkable structure of the world view of physics and also of biology will always be in the back of these considerations.

This view of the world is on the one hand a product of the human mind, and on the other hand it states that mind is itself a product of evolution, maybe only a marginal one.

4.2 Where Science and Religion Touch

What we mean by the word religion is an extremely many-layered conception, and its relation to science is similarly complex. I am concerned in this section only with some simple, but important aspects which by no means exhaust the topic.

In contrast to science, where the findings are considered at best as a kind of approximation to the true structure of the world, religion rests on the assumption that it possesses knowledge of absolute truth, and explanations are just derived from that basic belief. Apart from that there are differences between religions, and to be definite, we shall consider some fundamental elements of the Christian religion, such as the belief in one God, who has created the universe, and sustains it. In addition, Christians believe that God has a purpose, a plan for his creation, connected to the meaning of our lives, and that our real existence reaches beyond our world-immanent being.

Can science say something relevant regarding these points? In a very restricted sense it can: Whenever the real world is in focus, the findings of science have absolute validity. Although scientists are not dogmatic and always ready to examine their models critically, they know what is true and false within the realm of science. The question is not whether you like some idea or thought better than another one, but whether you can decide by scientific means whether it is right or wrong. If it can be decided, there is no room left for discussions. A law of nature just holds, you do not have the freedom to obey it or not. The scope of science is limited, however, and unfortunately only few questions can be treated so clearly.

There are other approaches to the world, but these philosophical or theological considerations cannot verify or falsify results of science. Thus our religious belief must be able to pass the tests of scientific reasoning.

4.2.1 Straightforward Influences

There are very simple interactions of physics and religion: Research leads to scientific explanations of natural phenomena, and the Gods, who have been necessary to understand lightning and thunder, rain, sunrise and sunset, and so on, are removed from the world. As the scientific investigation of the world proceeds, the ideas about the inner coherence of phenomena become more subtle, and are reduced to a few fundamental conceptions. This could be a process also working in religion, and leading to changes in religious teachings.

The physicists spend their time in bringing up new models, examining them in experiments and theoretical computations, and pursuing them further or discarding them. They form new ideas and conceptions as an essential element of their methodology. Would it be possible to make religion more open in that respect? I think one cannot have too radical a change, because there must be an absolute commitment in religious belief. But perhaps the images projected by religion to make the world understandable may be modernized a bit without damaging the essential substance.

4.2.2 The Story of Creation: In the Bible and in Modern Science

Let us try to look at the biblical tale of the creation of the world in the light of modern science. This text has already been reinterpreted by theologians, and it seems clear from modern biblical research that it should not be understood too literally. So let us try to give an interpretation taking into account results from science.

Here we may ask the simple question, how we can still speak of God as the creator of the world, if we take the findings of modern science into account. Clearly the biblical text does not aim at a scientific statement about the world, but rather it is a statement of faith assuring us that we owe our existence to God, who has created the world. But I think that we should not see it completely as a symbolic myth, because without any connection to the real world around us the work of a Creator would remain a noncommittal metaphor.

First of all, it is very remarkable that the modern cosmological big bang theory fits very nicely to the biblical statement that God has created the world from nothing. The cosmological model begins with the big bang, when everything, even space and time originates. Thus we can identify a moment of time in our past, when the world was created, but we cannot speak of a time before the big bang. If God has created the world, he must have done this according to the big bang theory outside of space and time. This conclusion is theologically certainly relevant, and it is not new at all: Already St. Augustine has pointed this out in his Confessions (Augustinus Confessiones lib. XI, Chap. 12):

"See, I answer to him, who asks: What did God, before he created heaven and earth? I do not give him the answer that somebody once gave jokingly, when he wanted to evade the difficulty of this question: "He prepared hells for those, who are keen enough to investigate these deep mysteries." ...But I call You, our God, the Creator of the whole Creation...Because it is precisely this Time which You have created, and there could pass no times, before You have created Time. If there were no time before heaven and earth, how can anybody ask what you did then? There was no "then," where there was no time." I can wholeheartedly subscribe to this view as a modern scientist.

The making of the first man from clay is certainly an adequate illustration of the creation for humans living in the two-stream country between Euphrates and Tigris with its abundance of clay. The understanding we have today would lead us to modify this view, and to put forward the interpretation that cosmic development and biological evolution on the Earth were the tools of the creator when he created the manifold of living beings and finally man.

The biblical report can be seen as an acceptable scientific explanation at those historic times, but we may also accept it as being in harmony with present-day scientific knowledge. This reinterpretation would certainly not concern the central message which is definitely meant to give meaning to our life by emphasizing its value, because we have been created after the image of the creator. We are subjects standing before one, single God and Creator, not a multitude of gods.

Our attempts to look at the contents of our belief and at the biblical teachings in a rational way, and to consider reasons speaking in favor of or against them, are contrary to the attitude which says that believing is a completely irrational decision without any connection to reasoning. This uncompromising view which is represented, e.g., by the German theologian Karl Barth is, as we have said already logically sound. It appears to me, however, quite unsatisfactory. Why has God supplied man with a self-conscious mind, and with the intellectual capacity to reason? Obviously, because he wants us to make use of it, and not to switch off our intellect at critical points.

Therefore I believe the religious confession that God has created the world must include the scientific insights we have of this world. Otherwise the words "Creator of Heaven and of Earth" would be empty shells of words without any real content, and with it the God of the Bible a pale shape, distant, and unreal.

Certainly the biblical story of creation uses insights of science of times long ago, but nevertheless expressing the understanding of the world at that time: The teachings of the Babylonians of the origin of the world use the image of the heavenly ocean, whose waters cannot fall down on the Earth, because the firmament prevents them from doing so. We should also keep in mind that 3,000 years ago in the land of the two streams a divine revelation talking of a big bang, quantum fields, and evolution could hardly have found its way into an adequately written account. Even the religious revelations had to express themselves in images which were understandable for human beings equipped with the knowledge of nature of those times.

But does not the evolution of the universe, the development from a simple beginning to a complex system, contradict the biblical account of a single act of creation? Not really, as we can see by considering an adequate illustration derived from our knowledge of physics: An act of creation which happens outside of the categories of space and time, can bring the complete space–time into being at once, such that the apparent historical course of things is only due to our view of it, as beings confined in space–time. For the timeless creator the whole history is present in one instant. Theologians have developed the idea of a continuously proceeding creation, a "creatio continuans." Their statements are in harmony with all the new scientific insights.

4.3 The End

But what about the cosmological final state? As time goes on, all the supplies of energy run down, and our beautiful, complex world which has come into being must perish again. We have already sketched the dull and uninspiring end of the world as it is predicted by the laws of physics: The final state consists of slowly evaporating black holes in an expanding cosmos which at the very end is just filled with long-wavelength radiation. We are not really enthusiastic about this vision of the future. It provokes the question why the Creator set up the world in such a grand scenario, let it evolve to great diversity, only to submerge everything in a meaningless dreary final state. I have two consolations: First of all there is a long time for life to enjoy the cosmos, and secondly the prediction is based on our present knowledge which is certainly not complete. We do not know the nature of dark energy, and we have no idea what the presence of conscious mind in the universe means for its future.

4.4 Metaphors from Physics

One difficulty about religious teachings is the way in which they often seem to contradict our intuition which has been formed from daily experience. There is a similar difficulty when we try to express the insights of physics in everyday language. Conflicts with common sense happen quite often. But the illustrative pictures are just an attempt to express the abstract knowledge we have in an understandable way. In that sense they may help to clarify certain difficult religious conceptions.

We have talked in detail already about time, and how it loses its absolute status, if it is created in the big bang and perishes in black holes.

Absolutely remarkable is the experimentally verified consequence of Einstein's theory of relativity that the passage of time for an observer depends on where he is, and how he moves. For a massless particle like a photon there is no passage of time, even if it propagates for billions of years from the source to the detector. The time of emission and the time, when it is received, are one and the same instant of proper time for the particle. This timeless existence is a property of all things which move with the speed of light.

This is not directly connected with religious statements, but the conceptions of time, of eternity, and of timelessness occur also in the Bible. Such sayings can be understood better, if we see similar ideas at work already in the world of physics.

There are many more scientific insights which run contrary to our personal intuition. Quantum mechanics describes phenomena in physics absolutely correctly, and in agreement with experiments, but we have not succeeded to gain an understanding of it in images of our experience. All the illustrative descriptions seem to be self-contradictory. Electrons exhibit properties of waves and of particles depending on the experimental setup. Are they particles or waves? They are something which is not exactly identified by either notion. This appears to us like an intrinsic contradiction of quantum mechanics. But this contradiction derives just from our intuition, our expectation which wants to see either a wave or a particle. The founder of quantum mechanics, the great Danish physicist Niels Bohr, has introduced the idea of "complementarity" for such phenomena which seem self-contradictory to common sense. According to Bohr quantum mechanical objects have this character of complementarity, i.e., they possess properties which contradict each other. Terms and concepts from everyday language cannot do justice to such objects.

4.4.1 Complementarity

The problems or antinomies one meets, if one desires to explore the relation between subjective experiences or between the subjective self and the objective "outer" world, might be approached in Bohr's way of thinking. It seems that our space–time perspective splits an individual into two parts – into the biological "machine" guided by the electric and chemical processes in the brain, and into the subjective being, who is convinced of the reality of her or his feelings, convinced of being an "I" that is not part of the objective world. Bohr's idea of complementarity might help us to think of these two separate aspects as the two complementary sides of the same whole, unique being.

I find it very likely that this idea can also help to understand religious teachings better. Is it not apparent that the deepest truth cannot be expressed in terms of everyday language, and that therefore paradoxical statements have to be accepted? Complementary sides of the same true sayings seem contradictory from our commonsense point of view ("credo quia absurdum").

The features of our world which run contrary to our intuition have been discovered in physics only with the revolutionary discoveries of the early twentieth century. Before, all phenomena and processes could be understood without difficulty within the framework of classical physics as the movements of small solid bodies and their strictly causal interactions. The contrast between religion and science developed from this mechanistic and materialistic view of the world. The claim to explain everything in this way did not seem to leave room for a freely acting being, let alone for an autonomous God as creator of the world. We leave aside earlier conflicts between church and science which had and have their origin in human stupidity which according to Einstein is infinite, unfortunately.

The ideas and suggestions outlined here would deserve a much longer chapter, but this would also require more research in theology. Without going into details, we can venture to say that theological concepts like "almighty God of Creation outside of space and time," "Eternity," "Hereafter" do not easily find an adequate representation in terms of everyday language. Attempts to tie them to our daily experience lead to contradictions. We can learn from modern physics that we have to live with such seeming contradictions, if we want to say deep truths about the world.

4.5 The Origin of Life

The question of the origin of life is at the center of scientific and theological interest, but I have set it aside in this book, because I wanted to concentrate the discussion on fundamental issues of the basic physics determining the structure of the world. Two well-known physicists have thought deeply about the origin of life, and published very recommendable articles about it: Erwin Schrödinger, one of the founders of quantum mechanics and Nobel prize winner of 1932, has documented a lecture series in the book *What is Life?*. Freeman Dyson, who was born in England, and now lives in the USA, has made essential contributions to quantum mechanics, and his considerations are published in an article with the title "Why is life so complicated?"

Both authors use basic arguments from physics to find plausible criteria which an elementary biological system must have so that it can react to the environment, can survive, and reproduce itself.

How does a ball of entangled molecules become a living system? How is a living system consisting of many such molecules different from a nonliving one? The "primeval soup" in which the appropriate reactions occur randomly is a possible way toward the origin of the complex molecular structures which serve as the basis for life. There is also the concept that the self-organization of molecules on the surfaces of crystals, observed in nanophysics, may be responsible for the origin of life. This question is intensely investigated at the moment. Maybe life was not cooked in the primeval soup, but barbecued on a hot plate.

In general scientists agree that the origin of life on the Earth can be understood as a sequence of chemical and physical processes, although perhaps right now not yet in every detail. Cosmic influences apparently are not essential, only processes on the Earth are of importance.

Darwin's theory of evolution defines the framework which helps us to understand how random mutations in the giant molecules of the DNS and natural selection have cooperated to create the great diversity of species on the Earth over a time of a few billion years. There are at present no serious scientists, who would doubt the general picture, even though not all the details have been clarified.

It seems that any direct cosmic influence on these processes is negligible, but there may well be an indirect one, because the boundary conditions for the biological evolution on the Earth are defined by the time scales which determine the formation, the development, and the end of normal stars like our Sun. A star burning hydrogen in its interior at an even rate remains a few billion years in this state, until the hydrogen in its interior is used up. Then it blows up tremendously and becomes a red giant. When the Sun will reach that stage, in about 5 billion years, it will engulf the Earth's orbit in its outer shell, and the biosphere of the Earth will be destroyed. After that the Sun will shrink to a size comparable to the Earth, and end its active existence as a white dwarf.

This lies far ahead in the future. Now, for about 5 billion years the Sun has provided a uniform stream of light and temperate heat for its planets, and it will do so for another 5 billion years. This then is the time span available for the evolution of life. At present it is apparently half-time, so to speak. It is an exciting question, how the biological evolution will continue, and how the development in the future will be influenced by human intelligence. Hopefully in a way that the human mind can reach its full potential. I am quite optimistic that our offspring will be able to solve the problem of the exploding Sun.

An essential aspect of the evolutionary scenario is the origin of the human mind. Some long time ago it turned out to be an advantage for living beings to grow a bigger brain, and gradually the evolution toward human beings began. Mind and consciousness were the equipment of man, who began to comprehend the world, and to change it. How long will that continue? What peaks of evolution will humanity climb up to?

Freeman Dyson has indulged in an amusing speculation in that context. He has asked himself whether intelligent beings might survive permanently in the universe. His answer is "yes," if the universe is expanding forever, and if the beings can adapt themselves to any change in environmental conditions, such as to survive in airless space. Then the whole Milky Way could be settled, and exploding Suns would lose their horror, because one could travel elsewhere in time. But finally, in the very distant future, all the stars will be extinguished. The sources of energy will have dried up, except for occasional bursts of radiation, when black holes collide. The further survival, according to Dyson, would depend on the ability to use these occasional events.

Intelligent civilizations of those future times would go into a kind of hibernation for most of the time, and only wake up to make use of the energy of such burst events. They could survive infinitely long in this way. Dyson imagines a universe with never ending activity and continuous evolution. There would be no end to interesting things happening. A very agreeable and optimistic view of things.

These arguments, albeit very speculative, hint at the great possibility that the biological scenario alone does not determine the future, but that human cultural achievements which are passed on from one generation to the next and which also grow in the process have their part in it.

Since the environment influences the selection, there is also an influence of human culture on the biological evolution. It makes a difference whether evolution starts out from a highly advanced culture or not. Such a view brightens a bit the dark picture which evolution provides for the individual: Personal talent and ability, careful education, and brilliant accomplishments do not count at all – only the passing on of the genetic material is of importance. But now the purely biological aspects are no longer dominant. Mind, consciousness whichever you want to call it, has appeared in this world, beholds and analyzes the whole situation, and exerts a decisive influence.

4.6 Consciousness

The mind which holds the whole world in its grasp is a great gift of nature for us. The universe with its billions of galaxies and its evolution of 14 billion years can be comprehended by mind. But to understand mind itself, to understand our own consciousness, and to put it in relation to the world is a difficult task. We know about activities of the brain like thinking about the world, producing emotions and feelings. All this belongs to our consciousness. Besides that there are many activities of our brain of which we do not have conscious knowledge like the control of our vital functions, digestion, heart beat, transport of oxygen in the blood. In our dreams images emerge from the subconscious, and come into our conscious awareness. At the time of ancient Greece humans still held their dreams to be as real as their experiences when they were awake. There is a pretty story from ancient China about the Dao sage Chuang-tzu illustrating this: Chuang-tzu had a dream, in which he was a butterfly fluttering happily from one flower to another. When he woke up, he sighed and said that now he did not know any longer was he Chuang-tzu, who had dreamt to be a butterfly, or was he a butterfly dreaming to be Chuang-tzu.

Today we look at the world more soberly than Chuang-tzu, and do not count the dreams of butterflies and our own among real things. By the communication with others, learning from their experiences and knowledge, we make sure of the contents of our consciousness which are common to everybody, and therefore deal with the real world, and we learn to separate those from private productions of our brain.

Our "self," the feeling of our innermost being, cannot be shared directly with others. We guess from our own experience that others also have a conscious self. Another story from China lights up this aspect very nicely: A famous painting shows two wise men – perhaps philosophers or astronomers – standing at a pond and watching the gold fish in the water. "Look at the fish in the water, how happy they are," says one of them. The other is doubtful: "How do you know that the fish are happy?" Then comes the reply: "How do you know that I do *not* know that the fish are happy?"

The subjective experience which appears to us as an essential part of our personal identity cannot be perceived directly in others. We just assume that it is analogous to our own. It is not easy to say whether it is identical, or what differences there are. It is certainly questionable that we know much about the subjective experiences of gold fish, probably much less than about those of our fellow human beings.

The brain is doubtlessly the material basis of our consciousness. Electric and chemical processes between nerve cells cause acts of consciousness. Are the neuronal processes different, when conscious acts arise from them? Biologists, especially brain researchers, spend a lot of effort investigating special functions of the brain which may be correlated with conscious activity. They have succeeded in relating the activity of specific areas of the brain with the feelings or willful acts of test persons. Thus even subjective feelings have become an object of research. The biologist Martin Heisenberg speaks of the "empirical subject" who belongs to the real world and who can be investigated by scientific methods. How does a person react, when certain areas in the brain are stimulated, what kind of emotions are found to be represented in the brain? The neuroscientists have already made great progress in this field. If you held the opinion that the empirical subject is not all, that there must be an existential subject who essentially is our personal self, then you would come into a difficult position: You must name a property that has been omitted in the investigation of the subject. But by giving a description the missing link is already made into an objective part of the world, and is set up for scientific research.

It is clearly a restriction that scientific research is confined to objective facts. Scientists ask only those questions which have an objective statement as answer. Other questions, perhaps equally important, are not asked at all. Thus the questions asked in biology aim at the function, the purpose. How does a feeling of pain arise? How does color vision work? The answer is a functional description of pain or color vision, but not pain or color vision itself. The insight, how something enters or leaves our consciousness, is not yet an answer to the question what consciousness is.

Why is this question so difficult? In the world view of physics we are participants as well as observers, who try to comprehend the whole universe. As a participant, I am the empirical subject, who is accessible by scientific research, as the preliminary end-point of a long chain of Darwinian evolution. As an observer, as an existential subject, I am conscious of this situation, and also self-conscious. The existential subject has no place in biology and physics, and one can, of course, deny that there is anything like that at all. Our conviction of the existence of a nonobjective "self" would then be a self-deception caused by intricate neuronal feedback processes.

Our conscious self defends itself against such an opinion. Martin Heisenberg says: "For every one of us the feelings, emotions we have, when we hear music or a poem, as well as all our thoughts and memories are real. They simply are there, just like trees, mountains, the Sun, and stars." In comparison the world of physics is a shadow world, almost only empty space, where systems of electrons and nucleons act. If I were able to analyze a fellow human – let us call him Tristan so that we become a bit more familiar with him – down to the finest details of his electron and nucleon configuration, I would find nothing else but such a complex system of elementary particles, although I may assume that he has similar feelings and thoughts as I do. But I would find no indication of a spontaneous action of free will, or the outburst of an emotion. In strictly deterministic sequence, or statistically varying, but according to strict laws of probability, a specific configuration of the nucleon system Tristan would determine the next one. Tristan could analyze me in the same way, and would find the same result. Then we could communicate this outcome to one another, and conclude that we both walk through life as automatic machines, our subjective feelings nothing but an illusion.

Here we have to accept a remarkable aspect: The world view of science, the objective description of the real world of our experience, does not include the architect of the whole picture. An observer, who has a conscious view of the whole, can himself not be in there.

Erwin Schrödinger has pointed out this feature of the world view of physics very clearly. He has emphasized that the assumption of a purely objective world is a simplification of the problem of obtaining knowledge about the world, an assumption which preliminarily excludes the subject, who wants to know, from the complex of things which need to be understood. Thus all qualities produced by the senses would be missing in the objective world view, which would be "colorless, cold, and silent." In vain one is looking for the interface, where mind would act on matter. The situation is even stranger, because, according to Schrödinger: "even though the world view is and remains for every one a construction of his mind, and apart from that has no existence of its own, mind remains a stranger in these images, has no place there, can nowhere be found" ("obwohl das Weltbild selber für jeden ein Gebilde seines Geistes ist und bleibt und außerdem überhaupt keine nachweisbare Existenz hat, bleibt doch der Geist in den Bildern ein Fremdling, er hat da keinen Platz, ist nirgends darin anzutreffen"). Schrödinger concludes that therefore any attempt to introduce subjective aspects into the world view of physics must lead to inconsistencies and contradictions. A complete view of the world should not exclude the experiences of the conscious subject, but integrate them into the picture. So far, there was no success in this approach.

Let us listen to another physicist: Freeman Dyson has laid down his views on God, the world, and man in a series of books. *Infinite in All Directions* (Penguin Books 1985) gives a clear account of his beliefs concerning these questions.

He comprehends the history of evolution as a steady increase of the influence of mind or consciousness. Mind is very patient, it has waited for 14 billion years until it has composed its first string quartet. We cannot even imagine how far mind will evolve, if it is given enough time. Dyson says that for him there is no difference between his idea of mind as a principle active in the world and God. If mind becomes so complex that we can no longer grasp its works then we call it God. For him mind exerts control over matter, and is active in the real world. At the same time it represents the innermost self of everybody. Dyson sees no problem in ascribing both these characteristic features to mind.

He goes on to speculate: "I see three different levels on which mind or conscience is present in the world. The first level on which mind manifests itself are elementary physical processes, as we see them, when we experiment with atoms in the laboratory. The second level is the human experience of self-consciousness. The third level is the universe as a whole."

Dyson explains that atoms in the laboratory behave rather like active agents and not like inert substances. They appear to choose in an unpredictable way between various possibilities, just as the laws of quantum mechanics allow.

The universe as a whole also appears remarkable with laws of nature which make it a hospitable place for the growth of mind.

He thus believes that atoms, human beings, and God have a share in mind, in different degree, but of the same kind. We, as human beings, stand halfway between the unpredictability of the atoms, and the unpredictability of God. Atoms are small parts of our mental apparatus, and we are small pieces of God's mental apparatus. Our consciousness, our mind, is able to receive signals from atoms and from God.

Dyson emphasizes: "I do not say that this personal theology is supported or proved by scientific insights, I just say that it is consistent."

These quotations show clearly that the connection between our innermost self, our self-consciousness, and the real outer world is difficult to analyze. Phenomena in the realm of mind like "self-consciousness" or "free will" are described on a level different from scientific statements. The language used has a long tradition in philosophy and theology, and the translation of the concepts into ideas useful in science can easily fail, since the perspective in the humanities and in science is quite different, and the same words usually are not congruent in their meaning. How can we ascribe something like "will" or "free will" to a system of atoms which evolves following definite laws of physics?

A symphony by Mozart could be represented by a physicist as a temporal sequence of fluctuations of the air pressure at the ear of the listener, but there is no doubt that something important would be lost in that procedure. On the other hand, in physics the sounds are nothing else but pressure waves in the air which reach our ears. In analogy we can say that consciousness and mind find their representation in the electric currents and chemical reactions in the neuron cells. But it is apparent that this representation is incomplete.

Schrödinger spells out very distinctly the antinomy between our subjective feelings, the conviction that we are able to act freely, and the molecular system which makes up myself according to science. This antinomy exists, because the investigation of the world in physics excludes the subject totally. If we postulate the existence of mind as a kind of substance existing outside of physical reality, it remains powerless without any possibility to act in the real world. Dyson circumvents this difficulty by postulating "mind" as an additional property of matter which is already present in atoms, and which takes control over matter as the complexity of the world increases.

We cannot but collect and order our experiences in space and time, and although we see the limitations of this world view, we are not sure whether there is something – such as mental phenomena – beyond these limits.

4.7 The Argument from Design

Ever since human beings have thought about the universe, they were preoccupied with the idea that the cosmos must have a creator. The "divine watchmaker" has constructed his world at the beginning of time like a precise clockwork, and now it is just running along. This popular picture demonstrates also the inherent weakness of this idea: The cosmic watchmaker, who set the universe in motion, must have a predecessor, a superwatchmaker, who has created him. An infinite hierarchy of watchmakers – without a beginning – is the natural consequence of this argument, cut off only, if one appeals to the existence of a "first watchmaker" in the sense of an Aristotelian "unmoved first mover."

The cosmic watchmaker is an illustration of the classical teleological argument for the existence of God (the picture has first been suggested by William Paley in the eighteenth century): The functionality and complexity of the world around us can only be understood, if some design is behind it. Darwin's theory of the evolution of the species by random mutations and selection has brought about the downfall of the teleological argument. Evolution proceeds without a purpose, and without a design. As a consequence it seems unnecessary to believe in God as the Creator of the world, and thus we all can be happy atheists, as prominent biologists are never tired of preaching.

In my opinion, we should be a little bit more careful. The fact that the teleological argument has been eradicated from physics does not mean that we have obtained a scientific proof of God's nonexistence: Absence of evidence is not evidence of absence! I agree that the teleological argument must be banned from physics, where we do not tolerate actions at a distance – neither in space nor in time. But the argument may have a place as a metaphysical consideration in philosophy and religion.

If we look at the evolution of structure in the cosmos, there are two quite obvious aspects: More and more complex systems come into being in the course of time, and the rich diversity of the world is ever-increasing.

The admirable richness in color and forms of plants and animals indicates that nature tries to create as many different forms as possible. Evolution is not geared toward minimalistic efficiency. It seems rather that everything that is possible is also tried out. We may conclude that a principle of "maximum diversity" is at work, as Freeman Dyson has said.

4.7.1 The Strong Anthropic Principle

These phenomena seem to suggest a plan or a purpose driving the world toward some final goal, and it can be very rewarding to speculate about such aspects. But within the frame of a scientific discussion, we must modestly describe the processes without asking for a purpose, design, or final goal. It is not allowed in science to look for the causes of certain phenomena from the point of view that a certain goal should be achieved. Thus we cannot argue that the constants of nature have their precisely determined values, because the cosmos should generate intelligent human beings.

We cannot deny that such arguments have a certain charm, especially if we consider the delicately balanced constants of nature. The smallest deviation of their values would lead to an essentially differently structured world which would not be able to sustain life of our kind.

We have extensively discussed the anthropic principle in its "weak" form in Chap. 2: We can conclude from the existence of intelligent beings that the universe has those properties which are consistent with the evolution of intelligent beings. This is nothing more than logical consistency, but it has merits as an illuminating indication of certain connections in the world which could not yet be explained by physics.

The "strong" anthropic principle makes the much stronger claim that the laws of nature are as they are because the possibility must be guaranteed that higher forms of life and finally man will evolve.

It is actually not a physical principle, but a religious one. Just replace the somewhat pale condition "possibility of the evolution of intelligent beings" by the interpretation that the extreme fine-tuning in the universe is a sign of God's creative activity directed at the evolution of man. Then the religious tinge of the strong anthropic principle becomes evident. There is nothing wrong with thinking religious thoughts, and the strong anthropic principle has motivated even some physicists to religious statements. There is only one strict requirement: Such arguments are not valid within the context of scientific reasoning.

We should tolerate the strong anthropic principle as an illuminating consideration in the realm of religion and metaphysics, but expel it from physics. It carries an antiphysical smell, in some sense, because the cosmic fine-tunings point to interesting scientific questions, but there is no answer from physics, only an anthropic explanation. Thus the search for a possible derivation of an answer within physics is not attempted, if one accepts that principle. Physicists should not throw the towel into the ring too quickly, but rather look at these arguments as an encouragement for further research.

While the strong anthropic principle sees man not as a marginal appearance in the cosmos, but as the ultimate goal of cosmic evolution, there is the adverse current of thought adhering to the idea of "parallel universes" or the "multiverse." The argument runs as follows: It is not a miracle that we find ourselves in a universe that seems to be finely tuned to our existence, because everything which is physically possible exists also in reality, but possibly in another universe, causally separated from ours. Among this manifold of worlds is also one, perfectly suited for us, the "best of all possible worlds."

One possible reason for the existence of many universes is seen in certain results of string theory: One expects the existence of many different ground states or vacua in this theory, because after compactification to a theory in four-dimensional space–time, there are many possible classical configurations attached to each space–time point. Each of these configurations – their number is estimated as 10^{500} – should have its own vacuum state, and evolve as a separate cosmos. Nobody knows whether our universe finds a corresponding configuration among these many mathematical solutions. The name "multiverse" does not reveal a lot of sensitivity for language, and the bizarre strategy of multiplying the number of universes ad libitum to evade some problem of the mathematical formulation of the theory raises suspicions.

4.7.2 Creationism

We must make a few remarks here on a fundamentalist belief which has many militant followers especially in the USA. The "Creationists," as they call themselves, demand that biblical revelation taken verbatim is the only source of truth, and is the absolute authority also on all questions of science. Scientific facts can only give testimony of the same truth as the Holy Scripture. This leads to bizarre statements regarding the evolution of life on the Earth like "all basic species were created directly by God in the first week of creation" or "creation happened about 6,000 years ago." Creationists see their main enemy in Darwin and his theory of evolution. They argue that it is a bad theory, because no macroevolution, and no transition between species has ever been observed. Also several well-observed species like the crocodiles have not evolved at all in the course of time. These are valid objections, and it is also legitimate to criticize a scientific theory. Two points should, however, be made very clear: First of all no scientist doubts that the principles of Darwinian evolutionary theory are correct, even if not all gaps in our knowledge have been closed. The idea that small genetic mutations sometimes happen in such a way that the mutant offspring have a slightly better adaption to the environment, and that therefore these mutations persist among numerous variations is almost self-evident. "Very probably the more probable happens" is the basic idea, and if you want to doubt this you have to doubt logic. Secondly, if creationism were true, nothing in cosmology, physics, and biology would make sense. So it is just that: nonsense. I myself find the idea distasteful that God should have faked historic evidence like fossils and radioactive elements, just to prove scientists wrong.

4.7.3 Intelligent Design

The proponents of intelligent design have revived the old teleological arguments in favor of the existence of God. They ask with their founder Phillip Johnson "Do we owe our existence to a blind materialistic process or to a purposeful creator?" They argue that intelligent causes are necessary to explain complex information-rich structures in biology, and they try to demonstrate such cases.

A system which needs all of its parts to become operational could not have evolved by numerous small changes of less complex precursors, they argue. Such an "irreducible complex" system would indeed be a good argument against Darwin's theory, but the search for such a system has been in vain so far. It is quite a hopeless task anyway, because even if ID followers succeeded in showing that Darwin's model was insufficient, other scientific models would be constructed which might solve the problem in a scientific way. In fact, if you look at biological evolution, you find little evidence for design. Organs are never optimized or replaced by well-designed new structures, but rather the old stuff is used again and again, repaired, or improved a little bit. When it works sufficiently well to get by with, it is left as it is with unused spare parts all over the place (see A. Kreiner (2008) for a more detailed discussion).

The attitude to look for gaps of knowledge in our present theories, and to propose these gaps as a refuge for God and for religious belief, is certainly not scientific, and it is not even intelligent. The clothes of this movement are scientific, but intelligent design is not science, it is a religious movement, and as such it does not belong in science classrooms. There is a fair chance that the gaps in our scientific knowledge will be filled by future research results, and religious belief residing in such gaps will be expelled, and end up homeless in the end. I believe that the option is not that we have to adhere to a strict naturalism and atheism, if there are no gaps in our explanations, but that we find truly room for religious belief, if we ask, where the ultimate laws come from which hold the world together and make it evolve.

4.7.4 Summary

What can we learn from our discussion of the argument from design? I think the following quotation (from A. Kreiner 2008) from George Coyne S.J. summarizes it perfectly: "If they respect the results of modern science, and indeed the best of modern biblical research, religious believers must move away from the notion of a dictator God or a designer God, a Newtonian God who made the universe as a watch that ticks along regularly."

4.8 The Origin of the Cosmos

Why is there a universe at all? Why does something exist and not nothing? Can these questions still be discussed in the realm of science? The big-bang model avoids these questions, and we must look for models which tell us a bit more about the beginning. Since we understand better and better how our existence on the Earth is connected with the cosmic evolution, we are also affected in our self-understanding by questions touching on the origin of the cosmos.

Quite understandably many cosmologists are deeply moved by that question. They find the suspicion that the picture of the big-bang model could be taken as an indication of a divine act of creation disagreeable, because they want cosmology to stay within the realm of physics. Therefore they keep constructing new variants of a theory of the origin of the cosmos. The simple big-bang model does not say anything about the true beginning of the world. All questions about the origin are pushed into the mysterious, singular beginning, where even space and time originated. As a plain physicists one would say that the occurrence of this singularity points to the limits of the theory. General relativity theory has the unique property that it has described its own downfall. Many attempts have been made to formulate a more complete world model. We have mentioned the suggestion by Andrei Linde, who assumes that the classical big bang model is preceded by a phase, where the universe begins in a completely chaotic way filled with fluctuating scalar fields of widely different magnitude. Every now and then a small region gains the right properties to undergo an inflationary phase, and grow rapidly by an enormous factor. Andrei Linde has painted more details of this picture: Again and again small regions with a possible inflationary evolution could arise from quantum fluctuations. Every one of these "inflation seeds" leads eventually to the start of a new "big-bang universe." Other regions stay in the phase of fluctuating fields. In this manner infinitely many cosmic regions would arise, parallel universes which would lose contact with each other rapidly, because of the exponential expansion during the inflationary phase. This blowing up of new worlds could be a continuous process without end or beginning. We live in one of the cosmic "bubbles," a universe well suited to living beings of our type, all others we could not contact. This might be at least a consistent picture of the beginning of the world. But, of course, this wild speculation needs a better mathematical formulation.

The argument by Roger Penrose (see Chap. 2) concerning the choice of the initial state of the cosmos is quite remarkable too. Our world has been selected as one of $10^{10^{120}}$ possible ones. This argument comes as close to a proof of the existence of God, as it seems possible within the realm of physics. It is, of course, not really a proof, because the actual existence of our world could be in the sense of probability estimates, a very improbable event. These statistical considerations can nevertheless inspire great wonder.

Recently many attempts have been made to postulate and investigate a kind of "pre-big-bang" structure within the context of string theory. The great appeal of these purely mathematical constructs, and the great claims that are made should not distract from the fact that these considerations have not made the connection with physics yet. Therefore these theories in their present status contribute little to our understanding of the world.

All these speculations about the beginning of the universe are uncertain explorations in a field which at present is beyond our knowledge of physics. In my opinion it would be best to classify the question of the origin of the cosmos as metaphysical, because it really cannot be answered within physics, where one is confined to follow and describe the processes within the universe, and within the cosmic history.

These speculative attempts to explain the origin of the universe show nevertheless that there are possibilities – mathematical constructions – to think in a rational way about the big bang, and maybe to find a reason in physics why it happened.

It seems somewhat strange that there is no motivation to look for further explanations, if it is assumed that the cosmos has been in existence for an infinitely long time. It seems easier to be content with that fact than with an origin in a big bang. But one might, of course, ask also in the case of an eternal universe, why there is something at all, and why the world seems to obey certain simple laws.

Here we have been led by the methods of physics to a boundary of our world view. Is it an absolute limit or a threshold beyond which even deeper insights into the interconnections of the world are waiting for us? A final, valid explanation for the existence of the cosmos can probably not be derived from the models of physics, because this problem touches without doubt metascientific areas.

The mental attitude inspired by the search for an absolutely true and final explanation of the existence of the cosmos is nicely illustrated by an anecdote of the Swiss physicist and Nobel prize winner Wolfgang Pauli: Pauli has died, arrives in heaven, and immediately wants to know the final explanations. How is all that set up, and how does it work – beginning of the universe and cosmic evolution? "Here is the blackboard, please explain!" says Pauli, and God hesitating a bit steps up to the blackboard and starts to write a formula. Pauli jumps up at once, grabs the sponge, wipes the formula off the blackboard, and shouts: "No! No! That's not the way! I have tried that already myself!"

4.9 A Principle of Creation

Can we find any argument for an orientation of the cosmic evolution, an arrow of evolution, just as we derive an arrow of time from the expansion of the cosmos within the realm of science? Biologists correctly emphasize that the evolution on the Earth is totally undirected, a random process that does not show any trace of being oriented toward, say, greater complexity of the organisms, or greater variety. It seems to me that they are afraid of running into the trap of teleological statements which they had fought so zealously during the nineteenth century. But to the eye of the friendly beholder, the history of life shows a tendency toward great diversity and increasing complexity. Genetic mutations may be completely random, but the selection imposed by the environment may have an orientation, certainly a global one derived from the boundary conditions set by the universe.

The cosmic argument, the "strong" anthropic principle, the idea of the multiverse can all be seen as an attempt to gain some understanding of why we are here, and where we come from, to find some complete, unified world view in a sense. These arguments do not remain within the boundaries of our present knowledge, but that is alright, since progress in understanding must involve the transgression of previously existing boundaries. They cannot of course be seen as statements of science, like Kepler's laws for instance. But we can accept them as metaphysical question marks inspiring further thought on these questions.

Let me insert at this point some thoughts on a principle of creation motivated from quantum physics and evolutionary theory. The picture I want to paint in the following is a generalization of Darwin's theory of the evolution of the species, also a generalization of the idea of the "self-organization of matter," enriched by quantum mechanical concepts. Although the line of arguments is close to scientific reasoning, it is a highly speculative sketch leading beyond science, but not without intellectual pleasure. (See the book by Peter Kafka in the bibliography.)

Darwin's theory describes the evolution of the species as a consequence of random changes in the genetic material (mutations) and subsequent selection by the environment of suitably adapted organisms, the "fight for survival." Mutations are "quantum jumps," i.e., random, macroscopically hardly noticeable changes in the giant molecules of the genetic material. Let us try to extrapolate this idea to more general quantum mechanical systems: As we have seen in Chap. 3, the quantum mechanical development of a system in time is not completely determined by the preceding states and the equations, as in classical physics, but only in a statistical sense. Starting from a certain state, the following step in time can lead to a variety of different states. The system "jumps" randomly to one of these states, not in a completely chaotic way, but with a probability determined by a given distribution of probabilities. When we observe the same state many times, it turns out that the states following in time occur randomly, but with different frequency corresponding to the probability distribution which is derived from the laws of physics. For an illustration let us look at the radioactive decay of uranium. The individual atom can decay immediately or after a few million years. It decays unpredictably and randomly. A certain mass of uranium which consists of many atoms, however, follows exactly the exponential law of radioactive decay.

A quantum mechanical system has an abstract space of potential states out of which it chooses randomly in the course of its space–time evolution.

Each single choice is the realization of a possible configuration in space and time. The result is determined by the probabilities assigned to the states in the space of all possible states (let us call it the space of potentialities), and by the actually realized states in the past.

If we look at the whole biological life on the Earth as one quantum mechanical system, then the individual random choice event is almost unnoticeable. The system is permanently "groping" for new states in the space of potentialities, and, if the new one can survive it is integrated as a building block which opens access to the realization of further new forms for the system.

Let us now in a bold speculation view the whole universe as a system of this type which explores the space of potentialities by groping around for accessible neighboring states. The space of potentialities is unbelievably rich and full of variety. It contains all the ideas of material and mental type, such as all mathematical structures, simply everything allowed by the fundamental laws. In this picture the real "creation" would be the creation of this space of potentialities. All configurations and forms in it are timeless, some will be chosen and realized by the system "universe" as it moves along in time. In this process the system as any quantum system experiences permanently small fluctuations and uses these to grope around and tap possible configurations which are so close to the actually realized state that they can be reached by a tiny change of the variables of the system, as e.g., the total energy, or the momentum of some part.

Shortly after the big bang, in the hot and uniform early phase of the universe there was only a very limited choice of neighboring states. The development was nearly deterministic. With continuing cosmic expansion and cooling, the number and complexity of realizable forms increased. Atomic nuclei, then atoms, finally galaxies, stars, and planets could arise.

Life on the Earth then evolved according to the same principle of creation which simply states: "Probably the probable will happen." The realization of the forms and shapes which are laid out as potential configurations happens randomly, but following a probability distribution which is derived from the basic equations. Different forms can be reached by the system with different probabilities. Among these forms there are some specially attractive ones which are approached by the system preferentially from different states by small fluctuations. This behavior is known from the physics of nonlinear processes. Complex patterns develop in the boundary area between fixed order and chaotic behavior, not completely stable, but existing for some time. In physics these states are called "strange attractors." They attract the system which tries to stay close to these attractors as in a state of equilibrium. Although the basic equations in principle also contain all the information about these attractors, they are in fact so complex that even in simple nonlinear systems they have been discovered only experimentally. The climate of the Earth is an attractor for the weather, health an attractor for the complex regulation mechanisms in the human body.

In this general picture there is reality, there are potentialities which can become "real," there are potentialities which have not yet been realized, can no longer, and can never be realized. Each state of the world realized at this moment can be viewed as a point in the immeasurably large space of potential states. The cosmic history from the big bang to the present is one line among many other possible ones in this space – maybe not even "the best of all possible." At each moment a state with relatively high probability is chosen and realized, as the cosmic history proceeds. The actual reality appears extremely improbable in view of the manifold of potential histories which have not been chosen. But that is not a problem at all – some real cosmic history had to happen after all.

If the assumptions which have led us to this picture are valid, then our principle of creation seems to be a simple logical consequence: "most probably the most probable happens." It is obviously the case in our world that more and more complex forms from the space of potentialities are reached by the system "Universe," and that it is not the case that random fluctuations lead back to simpler structures like a state of thermal equilibrium. This is on the one hand due to the boundary conditions – the permanent expansion and cooling of the cosmos allows to come into play successively weaker interactions and more intricate physical processes. On the other hand it seems to be true that more complex states in the space of potentialities have a relatively higher probability to be realized, and after their realization have more potential for further development. Such golden opportunities seem to have been accessed by the system, when it has chosen a path in the space of potentialities which has led to highly complex structures like the human brain.

I believe that the mental processes, our mind, our self-consciousness, and our emotions are not fundamentally different from other "real" processes as our universe traces out its path in the space of potentialities. Our thoughts and feelings wander around attractive ideas, and sometimes spontaneously they reach even more attractive ones on a higher level of complexity, but following the same principle that we observe in the development of matter to more and more refined molecules, and in the evolution of species toward an ever more complex biosphere. Everything which is called "real" by the scientists exists as an object in space and time, but this scientific reality is only a small selection from the empire of possible forms, one might even say, from the empire of spiritual forms in the sense of platonic ideas.

But does the matter as it is organized in a human body have the power to realize these immeasurable possibilities of mental structures? A simple example can serve to illustrate that this is indeed so. Let us estimate the number of possible links between nerve cells in a human brain: There are about 10 to a 100 billion cells, and each one of those is connected to about 10,000 others. The "information content" of such a system can be guessed, if we simply try to estimate the number of possible connections between points by drawing lines. For two points we can draw a line connecting them, or not draw a line. With three points, we can draw lines as follows: There are three different ways to draw one line, or two lines, again in three different ways, then three lines, or no line at all. Altogether we find eight possible connections. With four points we find 64 possibilities, with five points 1,024. How many points do we have to connect to get a number of links larger than the number of atoms in the universe? 24 are sufficient! The relations between nerve cells are so many that an immeasurable richness of mental structures can be exploited.

Would it not be a beautiful idea to interpret the individual person, the subjective self-consciousness of each human being, together with the material body as one attractor in the space of potentialities? The "body" existing within the boundaries of space and time tries to approach the individual person in a process of realization that develops in the course of time. Might there not be the chance to recognize the divine in the very foundation of all these attractors, all these attractive structures in the empire of mind? Does the insight not have a monotheistic ring to it that ONE principle of creation is at work – from the big bang to the structure of our brain, and to the human achievements of language and culture? That is certainly the limit up to which we can drive our speculations, as long as we want to base them on scientific knowledge. Our picture certainly extends beyond a purely naturalistic view, although we wish to explain everything as "natural."

It should be clearly realized that such speculations are in perfect harmony with religious beliefs. Creation as we view it is creation of the whole space of potential forms much richer than just a world in space and time. Much more magnificent is the Creator by this interpretation, and religious believers should be convinced that the path of our world in the space of potentialities, the freedom of choice inherent in it, is the will of the Creator not to determine everything in detail, but leave room for evolution, and even surprises.

4.10 Synopsis

The world we live in is stranger than one would think at a first glance. As soon as we look beyond our well-known surroundings, to the gigantic structures of the galaxy clusters, as well as into the tiny domains of the subatomic world, we find traits of the real outer world which pass beyond our daily experience, partly run contrary to our intuition, and challenge our imagination. Our "solid" everyday world is supported by an underground world of fields and particles which behave according to quantum physics in a remarkable nondeterministic way. The deep ideas of string theory finally trace everything back to the vibrations of small "strings," little packets of energy, completely immaterial and existing completely outside of the usual categories of space and time.

It is not clear at all, how this quantum world is related to the normal world of our experience. Therefore it is not yet possible to draw a coherent picture of the physical world, an intermediate, preliminary account must do.

Our solar system is situated in the outer regions of the Milky Way, a stellar system of about a hundred billion of stars. Billions of such huge stellar systems exist in the observable universe. And everything is the outcome of an evolution lasting for 14 billion years which in the beginning was no more than the uniform expansion of a hot gas. Besides matter and radiation, also space and time originated right at the beginning in the big bang. Here is an interface of cosmos and quantum world which we cannot yet describe with our present knowledge of physics. The evolution of the cosmos, a tiny instant of time after the big bang, is well understood, so we believe.

Our own existence is bound to a fantastic cosmic cycle: Every carbon or oxygen atom in our body comes from the interior of a star, where it was brewed under huge temperatures and pressures. The newly made atoms were scattered in space, and later used as building material when our solar system originated. Even beyond our own death these atoms will all be conserved. They will continue to participate in chemical reactions on the Earth. We are more than just the atoms we consist of. We are defined by the highly complex patterns in which they are arranged, the form in which matter is organized or organizes itself.

An essential feature of the cosmic processes seems to be an evolution from simple to complex structures. In the beginning there was only an almost structureless hot gas, but in the course of time a great variety of forms unfolded. The world seems to become the more interesting for us the more we know about it. Even in a very sober mood, we have to admit that the universe looks like a hospitable place for life. The fine-tuning of the constants of nature and the properties of the fundamental interactions make this possible. We know that as a fact, but we do not know why it is so. The question, why the cosmos is as it is, cannot be answered by scientific research. Only pseudoscientific arguments like the strong anthropic principle or the multiverse idea can make a contribution here.

Even if there are many open questions, it seems to be clear how the evolution proceeds in terms of physics: Strictly causal, or statistical laws of nature determine, how one state follows after another. In this world view there is no room for freedom, feelings, or belief.

But this strict and constraining explanation of the world is itself subject to strong constraints. If we follow the theories of physics to their ultimate consequences, we see that they have to be supplemented eventually: If space and time originate in the big bang and perish in black holes, then the order of the world in space and time cannot be everything. We are constructed such that we cannot but order our experiences in space and time, but our theories show that we need ideas that go beyond space and time to gain a complete understanding of the world. Surely we cannot know for certain, simply from thinking about it, that there is really something beyond space and time. But the deep analysis of physics removes obstacles which make it difficult for us to grasp such possibilities. It is quite astonishing that the theory exhibits itself the boundaries of its validity. If science cannot grasp the whole of reality, then the path is free for belief in a religious sense without the worries about the permanent conflict with the uncompromising results of science, or the contradictions to a simple deterministic-causal world view. This may not seem much, but it is remarkable, because it follows from scientific arguments. Surely this is true only for our present knowledge of physics, but in the future, more fundamental theories at which we guess now hint at a reality beyond space and time.

An important aspect in this connection is the effect of the method of science: All subjective feelings, emotions, even the conscious self are excluded, because of the constraint to give an objective description of the world. All this has no place in the objective, rational world view of science.

According to the current interpretation of quantum mechanics it seems, as if such an objective description could not be carried through. The conscious reaction of an observer seems to be necessary to fix the result of an experiment in reality. This Copenhagen interpretation of quantum mechanics seems, interestingly enough, to lead to a conflict with the ideas of brain research to understand consciousness as an objectively measurable activity of the brain. This field of research has had remarkable success in exploring the way our brain is functioning. The attempt to understand the whole world of the human mind, including the subject, convinced to be an irreducible self, from the objective description of the complex interactions between nerve cells in the brain is ongoing, but the outcome is still uncertain. We certainly can look forward to exciting times. Anyway, if the brain researchers succeed in their quest, no quantum mechanical system would ever enter into the real world, because to do that the conscious act of will of an observer is necessary, according to the Copenhagen interpretation of quantum mechanics. Both views cannot be right, it seems, because otherwise there would be no reality, and also no brain research.

For the time being the architecture of life has reached its highest level in the human mind, in the self-conscious being, who can reflect upon the world and his fate in it. Can evolution proceed any further? Perhaps the evolution leading to the creation of more and more complex systems is not yet finished? For mankind the biological evolution is probably no longer relevant, because we are no longer affected by natural selection. For us further development occurs through the cultural achievements of man which are passed on from generation to generation in a sense like "objective mind" in writings, recordings of sound and images. By that the brains of coming generations will be impregnated much more efficiently, and changed much faster than it would be possible by biological mutations and selection. Great potential and great freedom seem to be in store for us. Of course, the uncertainty remains that we see only the potential, but can never be sure. But as scientists we are used to live with insecurity.

The reduction to biological and physical phenomena does not tell us anything comforting about the role of mind in the world: The evolution of consciousness is a marginal event from the point of view of science, a random occurrence which could as well have not happened. At the end, when the stars will stop shining, and the galaxies will perish in gigantic black holes, then the world of mind will also perish and disappear. Although this sounds rather disappointing, it may not be the whole truth, because the picture of a cosmic evolution leading to more and more complex systems has been drafted by the human mind, is his world view. But it is a construction which excludes the architect, it leaves no room for a subjective conscious being. Maybe the picture can only be completed, if we find a way to reintroduce the creative mind into the world of physics.

Deep inside, I am convinced that the biological unit representing me, also includes an "I," my subjective innermost self, my soul – if you want to call it so – and is not at all a big self-deception, but a reality which points beyond the obvious biochemical existence. The biological and physical processes are the firm basis for life, and the foundation upon which mind can develop, but they are not everything. The evolution of the universe may be viewed as a continuously growing dominance of mind over matter, a very attractive picture of the cooperation of mental structures and the real world of physics. Whether this approach to overcome the clash between the deterministic material reality and the subjective confidence of a free will is convincing, may be left to the personal judgment of each of us. But such a sketch does not contradict scientific knowledge. We may even derive a cosmic moral law from it: We should do everything we can to preserve the diversity of life and of nature, such that the evolution of mind in the cosmos is not disturbed. We cannot even guess what great achievements such an evolution may attain.

We cannot substantiate such beliefs in terms of a scientific statement, neither can we derive religious teachings or contents of belief from it. If we take science seriously in its extreme consequences, we come to a gradual understanding of the fundamental building plan of the world. But the tension remains between our scientific knowledge confined in space and time and the longing we have for the deeper truth of our existence.

Bibliography

Cosmology (Chap. 2)

Börner, Gerhard: "Kosmologie", S. Fischer, Frankfurt am Main, 2002

Börner, Gerhard: "The Early Universe - Facts and Fiction", Springer, Heidelberg, 2003

Börner, Gerhard: "25 Jahre Kosmologie", in Spektrum der Wissenschaft, December 2003

Greene, Brian: "The Elegant Universe" (Vintage Books, N.Y. 2000)

Silk, Joseph: "The Big Bang" (Times Books, 2000)

Weinberg, Stephen: "The First Three Minutes" (Basic Books 1993)

Quantum World (Chap. 3)

Bruß, Dagmar: "Quanteninformation", S. Fischer, Frankfurt am Main, 2003

Davies, Paul James William, and Julia R. Brown (eds): "The Ghost in the Atom: a Discussion of the Mysteries of Quantum Physics", Cambridge University Press, 1986

Enzensberger, Hans Magnus: "Zugbrücke außer Betrieb" (http://www.mathe.tu-freiberg.de/ hebisch/cafe/zugbruecke)

Genz, Henning: "Elementarteilchen", S. Fischer, Frankfurt am Main, 2003

Heisenberg, Werner: "Schritte über Grenzen", Piper, München 1977

Schrödinger, Erwin: "Mind and Matter", Cambridge University Press 1967

G. Börner, *The Wondrous Universe*, Astronomers' Universe,
DOI 10.1007/978-3-642-20104-2,
© Springer-Verlag Berlin Heidelberg 2011

General (Chaps. 1 and 4)

Augustinus, Aurelius: "Confessions" book 11 (translation and Latin text by Otto F. Lachmann, Reclam, Leipzig 1888

Dyson, Freeman: "Infinite in All Directions", Penguin Books, London 1988

Heisenberg, Martin: in "Der Mensch und sein Gehirn: die Folgen der Evolution" (Meier Heinrich und Detlev Ploog (eds), Piper, München 1997

Kafka, Peter: "Gegen den Untergang - Schöpfungsprinzip und globale Beschleunigungskrise", Carl Hauser, München 1994

Kreiner, Armin: in "Gott denken und bezeugen", p. 542ff., Festschrift für Walter Kardinal Kasper (G. Augustin and K. Krämer (eds)), Freiburg 2008

Russell, Bertrand: "History of Western Philosophy", George Allen and Unwin Ltd., 1961

Schrödinger, Erwin: "What is Life?", Cambridge University Press 1967